Boundary Elements and other Mesh Reduction Methods XLIV

WITPRESS

WIT Press publishes leading books in Science and Technology.
Visit our website for the current list of titles.
www.witpress.com

WITeLibrary

Home of the Transactions of the Wessex Institute.
Papers published in this volume are archived in the WIT elibrary in volume 131 of WIT
Transactions on Engineering Sciences (ISSN 1743-3533).
The WIT electronic-library provides the international scientific community with immediate
and permanent access to individual papers presented at WIT conferences.
http://library.witpress.com

FORTY FOURTH INTERNATIONAL CONFERENCE ON BOUNDARY ELEMENTS AND OTHER MESH REDUCTION METHODS

BEM/MRM 44

CONFERENCE CHAIRMAN

Alexander H. D. Cheng
University of Mississippi, USA
Member of WIT Board of Directors

INTERNATIONAL SCIENTIFIC ADVISORY COMMITTEE

ORGANISED BY

Wessex Institute, UK
University of Mississippi, USA

SPONSORED BY

WIT Transactions on Engineering Sciences
International Journal of Computational Methods and Experimental Measurements

WIT Transactions

Wessex Institute
Ashurst Lodge, Ashurst
Southampton SO40 7AA, UK

K. **Dorow** Pacific Northwest National Laboratory, USA

W. **Dover** University College London, UK

C. **Dowlen** South Bank University, UK

J. P. **du Plessis** University of Stellenbosch, South Africa

R. **Duffell** University of Hertfordshire, UK

A. **Ebel** University of Cologne, Germany

V. **Echarri** University of Alicante, Spain

K. M. **Elawadly** Alexandria University, Egypt

D. **Elms** University of Canterbury, New Zealand

M. E. M **El-Sayed** Kettering University, USA

D. M. **Elsom** Oxford Brookes University, UK

F. **Erdogan** Lehigh University, USA

J. W. **Everett** Rowan University, USA

M. **Faghri** University of Rhode Island, USA

R. A. **Falconer** Cardiff University, UK

M. N. **Fardis** University of Patras, Greece

A. **Fayvisovich** Admiral Ushakov Maritime State University, Russia

H. J. S. **Fernando** Arizona State University, USA

W. F. **Florez-Escobar** Universidad Pontifica Bolivariana, South America

E. M. M. **Fonseca** Instituto Politécnico do Porto, Instituto Superior de Engenharia do Porto, Portugal

D. M. **Fraser** University of Cape Town, South Africa

G. **Gambolati** Universita di Padova, Italy

C. J. **Gantes** National Technical University of Athens, Greece

L. **Gaul** Universitat Stuttgart, Germany

N. **Georgantzis** Universitat Jaume I, Spain

L. M. C. **Godinho** University of Coimbra, Portugal

F. **Gomez** Universidad Politecnica de Valencia, Spain

A. **Gonzales Aviles** University of Alicante, Spain

D. **Goulias** University of Maryland, USA

K. G. **Goulias** Pennsylvania State University, USA

W. E. **Grant** Texas A & M University, USA

S. **Grilli** University of Rhode Island, USA

R. H. J. **Grimshaw** Loughborough University, UK

D. **Gross** Technische Hochschule Darmstadt, Germany

R. **Grundmann** Technische Universitat Dresden, Germany

O. T. **Gudmestad** University of Stavanger, Norway

R. C. **Gupta** National University of Singapore, Singapore

J. M. **Hale** University of Newcastle, UK

K. **Hameyer** Katholieke Universiteit Leuven, Belgium

C. **Hanke** Danish Technical University, Denmark

Y. **Hayashi** Nagoya University, Japan

L. **Haydock** Newage International Limited, UK

A. H. **Hendrickx** Free University of Brussels, Belgium

C. **Herman** John Hopkins University, USA

I. **Hideaki** Nagoya University, Japan

W. F. **Huebner** Southwest Research Institute, USA

M. Y. **Hussaini** Florida State University, USA

W. **Hutchinson** Edith Cowan University, Australia

T. H. **Hyde** University of Nottingham, UK

M. **Iguchi** Science University of Tokyo, Japan

L. **Int Panis** VITO Expertisecentrum IMS, Belgium

N. **Ishikawa** National Defence Academy, Japan

H. **Itoh** University of Nagoya, Japan

W. **Jager** Technical University of Dresden, Germany

Y. **Jaluria** Rutgers University, USA

D. R. H. **Jones** University of Cambridge, UK

N. **Jones** University of Liverpool, UK

D. **Kaliampakos** National Technical University of Athens, Greece

D. L. **Karabalis** University of Patras, Greece

A. **Karageorghis** University of Cyprus

T. **Katayama** Doshisha University, Japan

K. L. **Katsifarakis** Aristotle University of Thessaloniki, Greece

E. **Kausel** Massachusetts Institute of Technology, USA

H. **Kawashima** The University of Tokyo, Japan

B. A. **Kazimee** Washington State University, USA

F. **Khoshnaw** Koya University, Iraq

S. **Kim** University of Wisconsin-Madison, USA

D. **Kirkland** Nicholas Grimshaw & Partners Ltd, UK

E. **Kita** Nagoya University, Japan

A. S. **Kobayashi** University of Washington, USA

D. **Koga** Saga University, Japan

S. **Kotake** University of Tokyo, Japan

Boundary Elements and other Mesh Reduction Methods XLIV

Editor

Alexander H. D. Cheng
University of Mississippi, USA
Member of WIT Board of Directors

WITPRESS Southampton, Boston

Editor:

Alexander H. D. Cheng
University of Mississippi, USA
Member of WIT Board of Directors

Published by

WIT Press
Ashurst Lodge, Ashurst, Southampton, SO40 7AA, UK
Tel: 44 (0) 238 029 3223; Fax: 44 (0) 238 029 2853
E-Mail: witpress@witpress.com
http://www.witpress.com

For USA, Canada and Mexico

Computational Mechanics International Inc
25 Bridge Street, Billerica, MA 01821, USA
Tel: 978 667 5841; Fax: 978 667 7582
E-Mail: infousa@witpress.com
http://www.witpress.com

British Library Cataloguing-in-Publication Data

A Catalogue record for this book is available
from the British Library

ISBN: 978-1-78466-431-2
eISBN: 978-1-78466-432-9
ISSN: (print) 1746-4471
ISSN: (on-line) 1743-3533

The texts of the papers in this volume were set individually by the authors or under their
supervision. Only minor corrections to the text may have been carried out by the publisher.

Preface

This issue contains papers selected from the 44th International Conference on Boundary Elements and Other Mesh Reduction Methods (BEM/MRM 44). The conference series was founded by Professor Carlos Brebbia in 1978, with its first meeting held in Southampton, UK. For the next 44 years, scientists and engineers have used this gathering to exchange the progresses made in the field. The continued success of the meeting is a result of the strength of the research on boundary elements and mesh reduction techniques being carried out all over the world.

In the last few years, the conference has met with several challenges. One challenge was the passing of Professor Brebbia, on March 3, 2018. The conference lost its founder and the scientific world lost a giant. Professor Brebbia's legacy, however, continued. The 41st Conference was held in Southampton, UK, and 42nd in Coimbra, Portugal, with increasing number of participants.

In the year 2020, the world faced a new challenge – the Covid-19 pandemic. All in-person scientific gathering ceased during the year, which lasted into 2021. Both BEM/MRM 43 and 44 continued to hold, and were transformed into the online format. We persevered.

BEM/MRM 44, held in 15–16 June, 2021, was a successful conference. Although delegates could not meet "in person" in the traditional sense, they watched the presentation and met "face-to-face" on the computer screen through Zoom. The discussion sessions were lively, and colleagues were able to see and talk to each other after a long absence. We were able to accomplish the scientific goals set for the conference.

This volume collects some of the papers presented in the conference. The Editor would like to thank all authors for the quality of their papers and other colleagues for their help in reviewing the material.

The Editor
2021

Contents

SECTION 1
ADVANCED FORMULATIONS

ACCURACY ANALYSIS OF THE FAST MULTIPOLE METHOD FOR THREE-DIMENSIONAL BOUNDARY VALUE PROBLEMS BASED ON LAPLACE'S EQUATION

ANDRÉ BUCHAU
Institute of Smart Sensors, University of Stuttgart, Germany

ABSTRACT

The fast multipole method (FMM) is an established matrix compression technique for a fast and efficient solution of complex three-dimensional boundary value problems based on Laplace's equation using the boundary element method (BEM). Furthermore, the FMM significantly accelerates the post-processing of the solved BEM problem. The foundation of the FMM is a series expansion of Green's function into spherical harmonics. Both the accuracy and computational costs are controlled by the order of this truncated series expansion. The FMM has been successfully applied to a plethora of engineering problems from multiple disciplines and its overall accuracy has been impressively proved. The basic numerical properties of the truncated series expansions of the FMM have been already investigated in detail but for point sources only. Here, I revisit the fundamentals of the FMM and analyze the relevant properties of spherical harmonics, the error of the truncated series expansions of the FMM, the influence of coordinate transformations of these series expansions and potential discontinuities of computed field values at the interfaces between the cubes of the hierarchical octree scheme of the FMM. In contrast to earlier publications, I analyze and visualize the field values of distributed single-layer sources instead of point sources inside the total computational domain of an octree cube with the help of three-dimensional plots of their field values and their related approximation errors. For these studies, I put a focus on the accuracy of the reversed FMM algorithm, which is very well suited for meshfree post-processing and which is introduced here in detail for the first time.
Keywords: Laplace's equation, boundary element method, fast multipole method, spherical harmonics, post-processing.

1 INTRODUCTION

The fast multipole method (FMM) has been originally developed by Greengard and Rokhlin [1] to compute particle interactions efficiently. They have shown impressively that the computational costs for a system of N regularly distributed particles are reduced from $O(N^2)$ to $O(N \log N)$, if the well-known series expansion of $\frac{1}{4\pi|r-r'|}$ into spherical harmonics is applied in combination with a hierarchical octree scheme. They have applied a high order of the truncated series expansions to get accurate results due to the singular fields of the particles.

Nabors and White [2] have extended this approach to the solution of electrostatic field problems using an indirect formulation of the boundary element method (BEM). Later, an adaptive scheme for generalized BEM computations based on higher-order boundary elements and the Galerkin method has been introduced for electrostatic field problems [3]. In [4] it has been shown that the FMM competes very well with the adaptive cross approximation technique (ACA) and is furthermore applicable for a coupling of the BEM with the finite element method (FEM). In all these mentioned applications, the error caused by the FMM has been in the range of the total error of the resulting linear equation system and is barely visible in integral values like the capacitance coefficients of a multiconductor system [2].

WIT Transactions on Engineering Sciences, Vol 131, © 2021 WIT Press
www.witpress.com, ISSN 1743-3533 (on-line)
doi:10.2495/BE440011

Remarkably, despite the success of the FMM, a detailed analysis of the approximation error of the series expansions in the context of typical BEM integrals cannot be found in the literature. Hence, I present here a study of the approximation error of single-layer integrals as they occur in the context of indirect BEM formulations or as a part of the integrals of direct BEM formulations [5]. First, I revisit the properties of the used spherical harmonics and the multipole and local expansions of a single-layer source at a single boundary element. In contrast to particles, the scalar potential u stays finite at the boundary element. My study includes evaluations of the potential or its gradient, which is the field strength, in the related octree cube. Particularly, I put a focus on the interfaces between two cubes to investigate potential discontinuities of normally continuous field values caused by the truncated FMM series expansions. This point is especially important for precise meshfree post-processing as it is required for instance for the design of magnets for nuclear magnetic resonance (NMR) applications [6]. The reversed FMM algorithm, which was firstly applied in [7], is the basis for the following error analysis of the FMM transformations. There, efficient transformation operators as presented in [8] are employed for accurate and fast transformations of the multipole and local coefficients. Finally, I draw my conclusions based on three-dimensional plots of the fields and their approximation errors to first show the numerical properties of the multipole and local expansions but second to support a reasonable use of the FMM in the context of the BEM or related applications.

2   MULTIPOLE AND LOCAL EXPANSIONS

The starting point of my study is Laplace's equation

$$\Delta u = 0, \tag{1}$$

which is the fundamental partial differential equation of an electrostatic field problem based on the scalar electric potential u. The related Green's function for three-dimensional open space field problems is then

$$G(\boldsymbol{r},\boldsymbol{r}') = \frac{1}{4\pi}\frac{1}{|\boldsymbol{r}-\boldsymbol{r}'|}. \tag{2}$$

The point $\boldsymbol{r}'$ describes a point on the boundary elements or the single-layer sources, respectively. The field values are computed at the evaluation point $\boldsymbol{r}$.

The basis of the FMM is the series expansion of eqn (2) into spherical harmonics Y_n^m [9]

$$\frac{1}{4\pi}\frac{1}{|\boldsymbol{r}-\boldsymbol{r}'|} = \begin{cases} \frac{1}{4\pi}\sum_{n=0}^{\infty}\sum_{m=-n}^{n}\frac{r^n}{r'^{n+1}}Y_n^m(\theta,\phi)Y_n^{-m}(\theta',\phi') & r < r' \\ \frac{1}{4\pi}\sum_{n=0}^{\infty}\sum_{m=-n}^{n}\frac{r'^n}{r^{n+1}}Y_n^m(\theta,\phi)Y_n^{-m}(\theta',\phi') & r' < r \end{cases}. \tag{3}$$

r, θ and ϕ are the spherical coordinates of $\boldsymbol{r}$ and r', θ' and ϕ' are the spherical coordinates of $\boldsymbol{r}'$, respectively. Note, that the imaginary part of the double sum in eqn (3) vanishes.

2.1 Normalized spherical harmonics

Depending on the intended application of spherical harmonics, different normalization approaches are used. Here, I use the normalization which has been introduced in [1], since it is very advantageous in the context of the FMM regarding computational costs

$$Y_n^m(\theta, \phi) = \sqrt{\frac{(n-|m|)!}{(n+|m|)!}} P_n^{|m|}(\cos\theta)e^{jm\phi}. \tag{4}$$

n is the degree of the spherical harmonics and m is its order with $|m| \leq n$. $P_n^m(\cos\theta)$ are the associated Legendre functions and are computed using recurrence relations [9], [10].

The value of $Y_0^0(\theta, \phi)$ is constantly one. With increasing degree n, the number of lobes in a three-dimensional plot of $Y_n^m(\theta, \phi)$ increases, too. The direction of these lobes is adjusted by the order m. This is exemplarily depicted in Fig. 1 for the real part of $Y_1^1(\theta, \phi)$ and the imaginary part of $Y_4^2(\theta, \phi)$. The values of $Y_n^m(\theta, \phi)$ are additionally weighted with the factor r^n in the series expansions in eqn (3). That means higher order spherical harmonics are especially relevant for large distances to the origin of the coordinate system (Fig. 2).

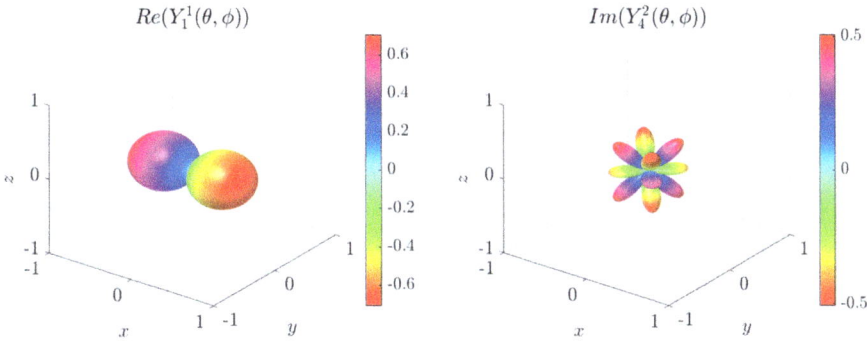

Figure 1: The real part of $Y_1^1(\theta, \phi)$ and the imaginary part of $Y_4^2(\theta, \phi)$.

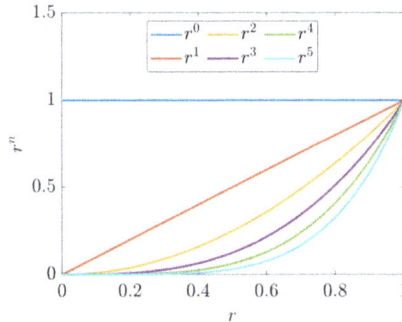

Figure 2: Values of the weighting factor r^n within the normalized interval of $0 \leq r \leq 1$.

2.2 Multipole expansion

The truncated multipole expansion of order L is obtained from eqn (3) for the case $r' < r$

$$u(r, \theta, \phi) = \frac{1}{4\pi\epsilon_0} \sum_{n=0}^{L} \sum_{m=-n}^{n} \frac{Y_n^m(\theta, \phi)}{r^{n+1}} M_n^m. \tag{5}$$

ϵ_0 is the electric permittivity of vacuum and u the potential of an electrostatic field problem.

The single-layer charge density σ on boundary elements inside a spherical domain around the origin of the coordinate system is summarized by the complex multipole coefficients

$$M_n^m = \int_A \sigma(r,\theta,\phi) r'^n Y_n^{-m}(\theta',\phi') dA'. \tag{6}$$

The M_n^m correspond to point sources of degree n at the origin and are equivalent to a monopole, a dipole, a quadrupole and multipoles of higher degree [11]. A discussion of the accuracy of eqn (5) follows in Section 3 in the context of the reversed FMM algorithm.

2.3 Local expansion

The local expansion is the evaluation of eqn (3) inside a domain around the origin of a spherical coordinate system, which is for the case $r < r'$

$$u(r,\theta,\phi) = \frac{1}{4\pi\epsilon_0} \sum_{n=0}^{L} \sum_{m=-n}^{n} r^n Y_n^m(\theta,\phi) L_n^m \tag{7}$$

with the complex local coefficients

$$L_n^m = \int_A \sigma(r,\theta,\phi) \frac{Y_n^{-m}(\theta',\phi')}{r'^{n+1}} dA'. \tag{8}$$

The L_n^m are not computed directly using eqn (8) but with transformations of the FMM algorithm indirectly with the help of the M_n^m.

Eqn (7) corresponds to a Taylor series expansion of the potential u at the origin of the spherical coordinate system. The degree $n = 0$ term represented by L_0^0 is equivalent to the potential at the origin and results in a constant potential inside the complete octree cube with its center at the origin of the spherical coordinate system. The three $n = 1$ terms add linear terms to the function of u in the considered domain. The value and sign of L_1^0 controls the linear dependency of u in the z-direction, the real part of L_1^1 controls the linear dependency on the x-coordinate and the imaginary part of L_1^1 adds a linear term dependent on the y-coordinate (Fig. 3). With the help of the higher order terms $n \geq 2$, the function $u(x,y,z)$ inside the valid domain is further shaped and adapted to the original function of u. In Fig. 4 the results of the imaginary part of L_3^3 are shown. There, the potential u in the FMM cube depends only on the x- and y-coordinates and is independent of the z-coordinate.

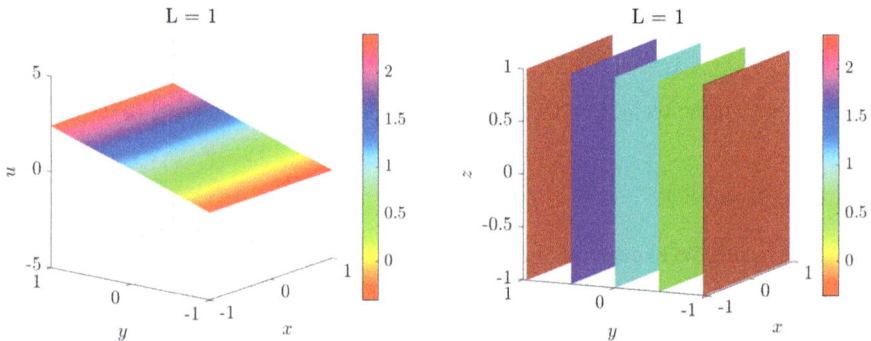

Figure 3: Potential u in a horizontal slice and its isosurfaces controlled by $\mathrm{Im}(L_1^1)$.

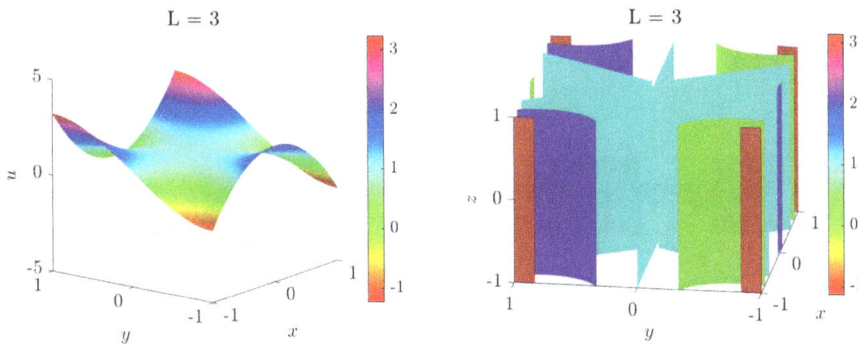

Figure 4: Potential u in a horizontal slice and its isosurfaces controlled by $\text{Im}(L_3^3)$.

It is worth mentioning that the product of $r^n Y_n^m(\theta, \phi)$ results in classical polynomial functions in Cartesian coordinates and is therefore well suited for an accurate approximation of scalar potentials $u(x, y, z)$, if there are no sources inside the considered evaluation domain.

3 REVERSED FMM ALGORITHM

The reversed FMM algorithm is perfect for post-processing and has been already successfully applied to the meshfree computation of field lines [7]. There, the idea that an FMM algorithm in reversed order is advantageous has been introduced in principle. Here, I present the reversed FMM algorithm in detail for the first time and analyze its numerical properties concerning the accuracy of its related series expansions and transformation operations.

3.1 Post-processing based on the reversed FMM algorithm

It is obvious to apply the FMM both to compress the matrix of the linear system of equations of the BEM and to improve its post-processing overall efficiency. However, there is a major difference between matrix compression and meshfree post-processing. In the first case, the position of all sources which are represented by the boundary elements is fixed and the field values are computed at fixed points, too [3]. Whereas an important property of meshfree post-processing is that only the sources and their positions are known in advance and the evaluation points are determined during the post-processing [12]. That means the structure of the underlying octree scheme is changing during the post-processing and the focus moves there especially to an efficient evaluation of eqn (7) including a suitable computation of the L_n^m.

All steps of the reversed FMM algorithm are summarized schematically in Fig. 5. Each step has an equivalent step in the original FMM algorithm [1] but I reversed the order of the steps and added conditions for the execution of a step to accelerate meshfree post-processing.

The reversed FMM algorithm is initialized with an octree that includes all boundary elements as sources. Then, in *step 1* the current evaluation point $\boldsymbol{r}_{ep}$, e.g. the next point of a field line, is added to the octree and its related cube C_{ep} is created if necessary. C_{ep} is identical to the cube in which the local expansion of *step 2* must be evaluated. To compute field values based on eqn (7), the L_n^m of C_{ep} are needed. If the L_n^m have been already computed within a previous step, e.g. for the preceding point of the field line, these L_n^m are taken in *step 3* and are evaluated in *step 4* using eqn (7) or the derivative of eqn (7) with respect to the global Cartesian coordinates to determine flux densities, e.g. the electric field $\boldsymbol{E} = -\nabla u$.

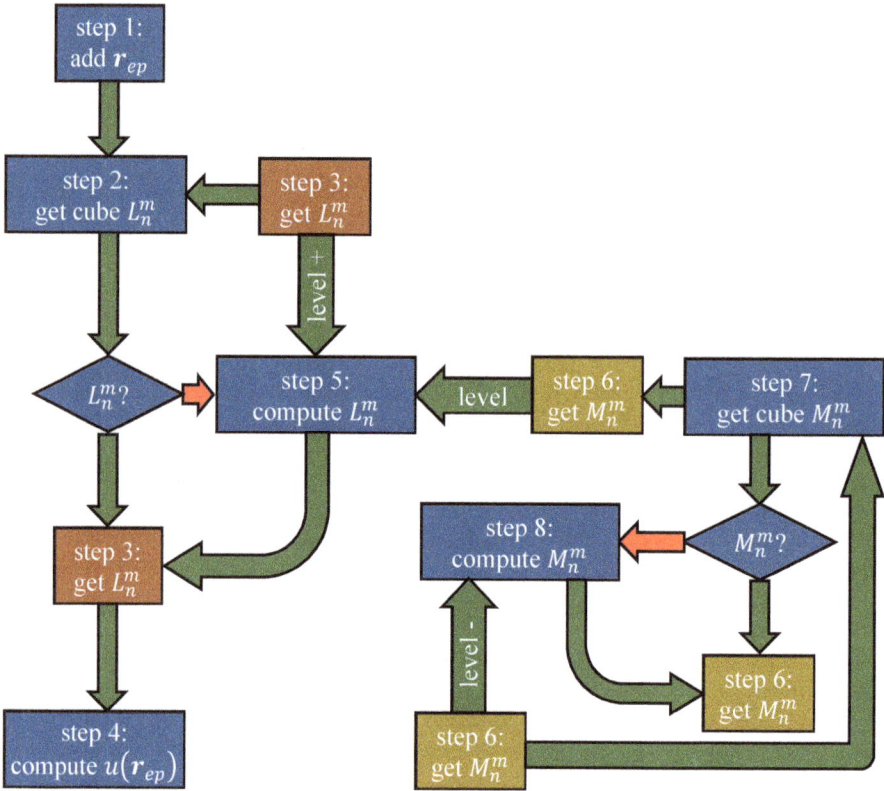

Figure 5: Computational steps of the reversed FMM algorithm.

Step 5 is only executed if the L_n^m have not been computed within an earlier computation. The L_n^m are computed from the related coefficients of the parent cube $C_{ep,p}$ of C_{ep} at the next coarser level of the octree (level+). This is equivalent to step 3 which now redirects to step 2 but using $C_{ep,p}$ instead of C_{ep}. If this loop ends in step 5, the loop is repeated if necessary. Finally, step 3 is reached and the $\bar{L}_n^m$ of $C_{ep,p}$ are transformed into L_n^m of C_{ep}

$$L_n^m = \sum_{k=n}^{L} \bar{L}_n^m T_{l2l}(n, m, k) \tag{9}$$

after a suitable rotation of the coordinate system to accelerate the computations [8]. The transformation coefficients $T_{l2l}(n, m, k)$ are computed once for each possible direction and modified for each octree level

$$T_{l2l}(n, m, k) = T_{l2l}^{direction} T_{l2l}^{root} T_{l2l}^{level}. \tag{10}$$

Additionally, the multipole coefficients M_n^m of far-field cubes at the same level are determined (*step 6*) and added to the local coefficients of C_{ep}

$$L_n^m = \sum_{k=m}^{L} M_k^m T_{m2l}(n, m, k). \tag{11}$$

Here, a rotation of the coordinate system is done first, too [8].

In *step 7*, cubes at the same octree level as C_{ep} must be found which are far away from C_{ep} to ensure the validity of the series expansions and which are not in a region that has been already considered in the far-field at a coarser octree level. For each relevant far-field cube C_{ff} it is checked whether the M_n^m have been computed (continue with step 6 and step 5) or must be determined in the current step (step 8).

The goal of *step 8* is to replace sources on boundary elements with sets of M_n^m. If C_{ff} has child cubes at finer octree levels, the $\bar{M}_n^m$ of the child cube are transformed using again a rotation of the coordinate system to accelerate the computations

$$M_n^m = \sum_{k=0}^n \bar{M}_{n-k}^m T_{m2m}(n, m, k).$$ (12)

At the finest octree level, eqn (6) is applied to compute the M_n^m directly from the sources on the boundary elements.

In the first run, the reversed FMM algorithm has computational costs compared to the classical FMM algorithm using fast transformations of series expansions based on a rotation of the coordinate system [8]. However, during the post-processing, the FMM algorithm is executed multiple times, e.g. for each point of a field line. Since the reversed FMM algorithm starts at the evaluation point r_{ep}, steps 1 to 4 are executed in most cases, only. In some cases, step 5 is additionally required. Steps 6 and 7 are needed only during the first run of the reversed FMM algorithm. Hence, the computational costs are in total significantly reduced by my reversed FMM algorithm compared to the classical FMM algorithm for post-processing.

3.2 Numerical properties of the local expansion

As I have shown in Section 2.3, the local expansion is very well suited to approximate scalar functions within a limited domain using appropriately chosen complex series coefficients L_n^m. Here, I apply the truncated local expansion (7) to the test configuration depicted in Fig. 6. A linear triangular Lagrange boundary element is used exemplarily to describe single-layer sources, here a constant surface charge density of $10 \ \text{nC/m}^2$. The cube structure of the applied octree scheme is shown by the black lines. The part of the boundary element, which is within its assigned cube, is highlighted using green color and its parts, which are outside its cube, are marked with red color. So, I study a typical situation where the octree scheme does not perfectly fit the boundary element mesh of the problem. The field of the source element is evaluated at the surface and inside the two cubes, for which in Fig. 6 the scalar potential is shown using the depicted color scheme. Since a linear problem is studied, geometrical dimensions do not matter and a normalized edge length of the depicted cubes of 1 m is used.

As above shown, the FMM algorithm or the reversed FMM algorithm is complex, and the approximation error of truncated series expansions and transformations is composed of errors of many different numerical operations. Hence, the total error of the FMM is difficult to specify and measure.

Here, I start the accuracy analysis of the FMM with studies of the accuracy of eqn (7). To exclude all errors of the reversed FMM algorithm, I compute the L_n^m directly by an evaluation of eqn (8) for the test element and with the origin of the local spherical coordinate systems of the cubes at the centers of the considered cubes. The potential u and the electric field $E = -\nabla u$ are evaluated in a slice going through the cubes and on the cubes' surfaces. These results represent the accuracy limits of the FMM for a given order L.

electric potential u

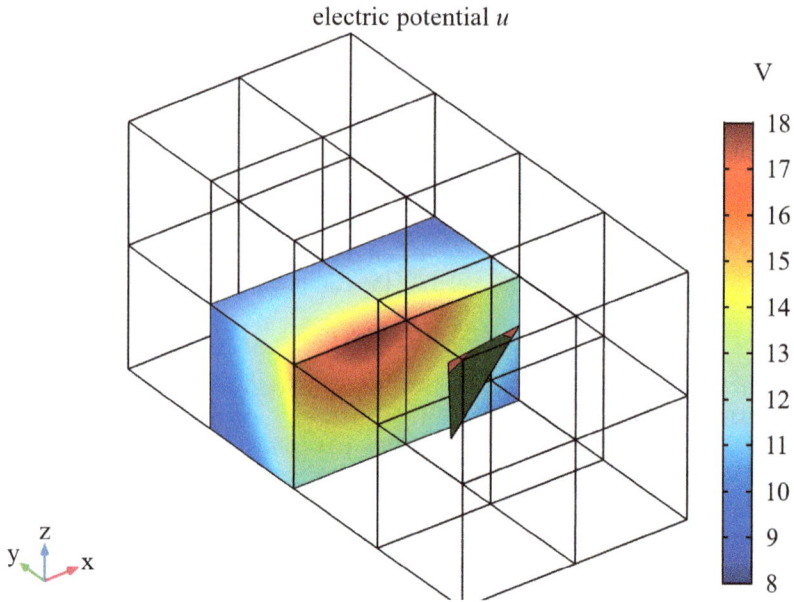

Figure 6: Test configuration for the presented accuracy analysis.

The potential u at the surfaces of the two cubes using eqns (7) and (8) with a truncated order $L = 5$ is depicted in Fig. 7. Furthermore, the relative error in comparison to the direct evaluation of the related BEM integral is shown. Even for a low order L, the color plot of u is smooth and a discontinuity of u at the cubes' interface is barely visible. However, the analysis of the related relative error leads to the conclusion that the discontinuity of u is not negligible. This is clearly illustrated by the plot of u in a slice in Fig. 8.

Of course, the relative error decreases if L or the distance between the element and the considered cubes is increased. That means it is in principle possible to find a good compromise between accuracy and computational costs for a reasonable approximation of u. Then, the approximation error is in the range of accuracy of the BEM and discontinuities of u at the cubes' interfaces are in practice negligible.

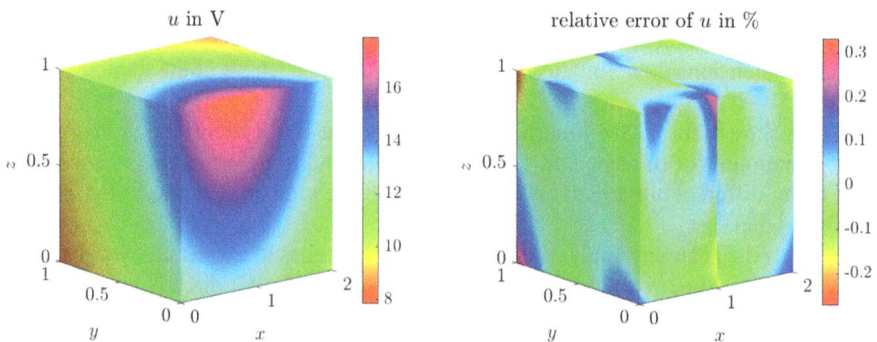

Figure 7: Potential u computed at the cubes' surfaces using eqns (7) and (8) with $L = 5$ and its relative error in comparison to direct evaluation of the BEM integral.

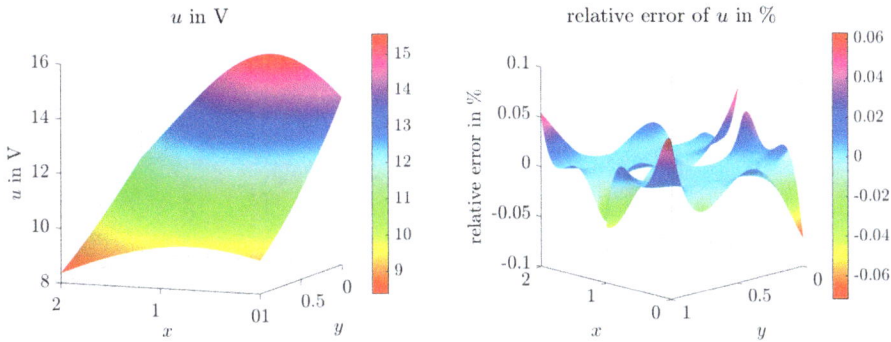

Figure 8:   Potential u computed at a horizontal slice using eqns (7) and (8) with $L = 5$ and its relative error in comparison to direct evaluation of the BEM integral.

In most applications, not only the potential u but also the electric field $\mathbf{E} = -\nabla u$ is of large interest for an engineer. Hence, the norm $E = \|\mathbf{E}\|$ and its relative error in comparison to the direct evaluation of the related BEM integrals is depicted in Fig. 9 at a horizontal slice through the considered cubes for an order $L = 5$. The plot of E shows a significant discontinuity and the relative error is much larger than the one of u. That means for precise field computations based on the derivative of the potential, a high order of the series expansions is required. Especially a visible discontinuity at the cubes' interfaces should be avoided.

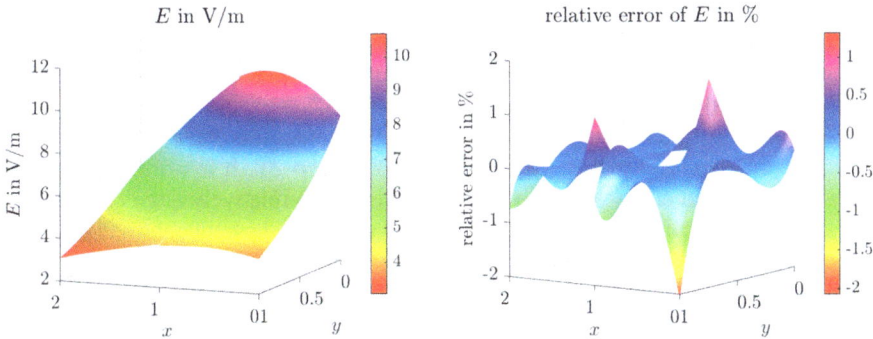

Figure 9:   Electric field E computed at a horizontal slice using the gradient of eqns (7) and (8) with $L = 5$ and its relative error in comparison to direct evaluation of the BEM integral.

I have repeated the computations of u and E with an order of $L = 10$ which gives in total satisfactory results for most applications. The relative error of both values is depicted in Fig. 10. For the studied configuration, the approximation error of u is negligibly small and even the approximation error of E has no relevant effect on the total results. Note, the approximation error can be larger if the source element has larger parts outside the related cube. Such situations often appear in the context of highly adaptive meshes.

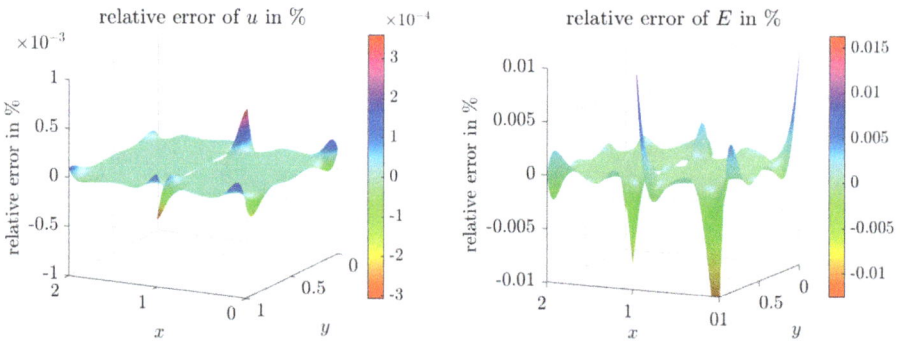

Figure 10: The relative error of u and E at a horizontal slice using eqn (7) and its gradient with $L = 10$ in comparison to direct evaluation of the BEM integral.

In total, the local expansion results in precisely computed field values if the order of the truncated series expansion is carefully chosen. Discontinuities at cubes' interfaces are then small enough not to be relevant in practice.

3.3 Numerical properties of the local-to-local transformation

In the preceding section, I have shown the potential and the superb properties of the local expansion for the fast multipole method and related applications. Here, I put the focus on the local-to-local transformation (9) and its accuracy. Particularly, I want to answer the questions of how this transformation influences the total accuracy of the FMM and which parameters can be used to improve the accuracy of the FMM.

In this study, I compute the local coefficients $\tilde{L}_n^m$ of the parent cube of the two considered cubes directly using eqn (8). The relative error of the potential evaluated using eqn (7) with $\tilde{L}_n^m$ in comparison to the direct computation of the boundary integral is depicted in Fig. 11 on the left side. The order of the truncated series expansion is $L = 10$. It is worth to mention, that the error at points that are close to the boundary element is relatively large. The reason is that in this test configuration the distance between the considered parent cube and the cube with the boundary element is smaller than usual in the FMM. At points in typical distance to the source element, the error is much smaller, here the points near the backside of the two considered cubes. That means in practice, the distance between the cubes which are used for the following multipole-to-local transformation and the part of the elements which sticks out of the related cube are the most important parameters to adjust the total accuracy of the FMM. The relative error of the potential using the local coefficients of the considered cubes after the local-to-local transformation is shown in Fig. 11 on the right side. This error is nearly identical to the previous one. A comparison of the potential values in both cases underlines this observation.

My conclusions of this study are that the local-to-local transformation has almost no influence on the accuracy of the FMM and the accuracy of the truncated local expansion is set by the parameters of the parent cube and not by the considered cube. The reason is that the potential inside the parent cube is described with the help of its local expansion and this series expansion is used to determine the coefficients of the local expansion of the child cubes. So to speak the transformed local expansion of the child cube is nothing else than the local expansion of the parent cube but with the shifted origin of the local coordinate system.

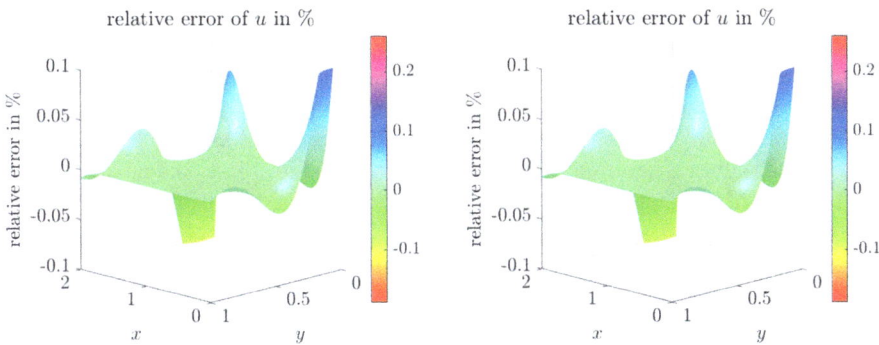

Figure 11:   Relative error of the potential of the local expansion of the parent cube (left side) and the transformed local expansions (right side) both for $L = 10$.

3.4 Numerical properties of the multipole expansion

The computation of the multipole coefficients using eqn (6) is the essential step of the FMM to convert the continuous charge distribution on the elements into multipoles up to the defined order L. The accuracy of the potential computed based on eqn (5) depends both on the number of multipoles and the distance between the region of the original charges and the evaluation point r. Within the FMM algorithm, the multipole coefficients of a cube are transformed using the multipole-to-multipole expansion into the multipole coefficients of the related parent cube. There, the same order of multipoles is available. That means the accuracy of the multipole expansion of the parent cube is comparable to the accuracy of the multipole expansion of its child cube. Analog to the local-to-local expansion the size of the domain of the sources is changed during this transformation and the overall accuracy is decreased.

I test this situation using the setup of Fig. 6. First, I compute the multipole expansion of the cube of the boundary element and evaluate the potential inside the two evaluation cubes (Fig. 12, left side). Then, I compute the multipole expansion of the parent cube both directly using eqn (6) and via the multipole-to-multipole transformation. Again, the shift of the coordinate system does not influence the accuracy. However, the enlarged size of the source domain lowers the accuracy of the potential close to the source domain (Fig. 12 right side).

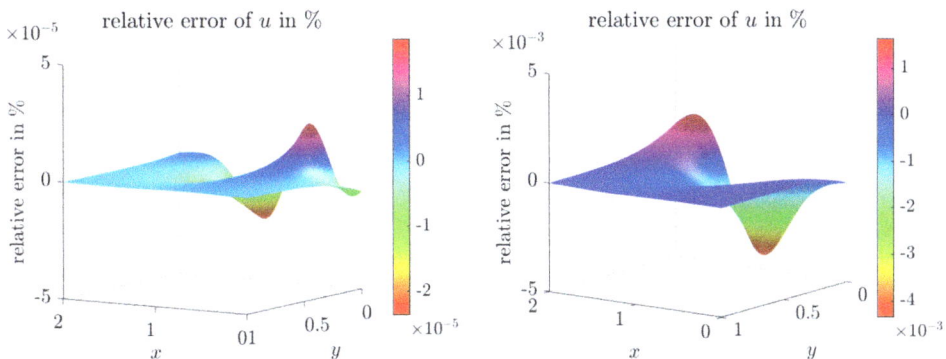

Figure 12:   Relative error of the multipole expansion of the element cube (left side) and the transformed multipole expansion of the parent cube (right side) both for $L = 10$.

3.5 Numerical properties of the multipole-to-local transformation

In my last study, I compute the multipole coefficients M_n^m of the cube of the source element using eqn (6). Then, I apply the multipole-to-local transformation to get the local coefficients L_n^m of the two evaluation cubes. The relative error of the potential u using the related local expansions in the cubes is depicted for $L = 5$ and $L = 10$ in Fig. 13. A comparison with Figs 8 and 10 shows that the multipole-to-local transformation has no visible influence on the accuracy of the FMM.

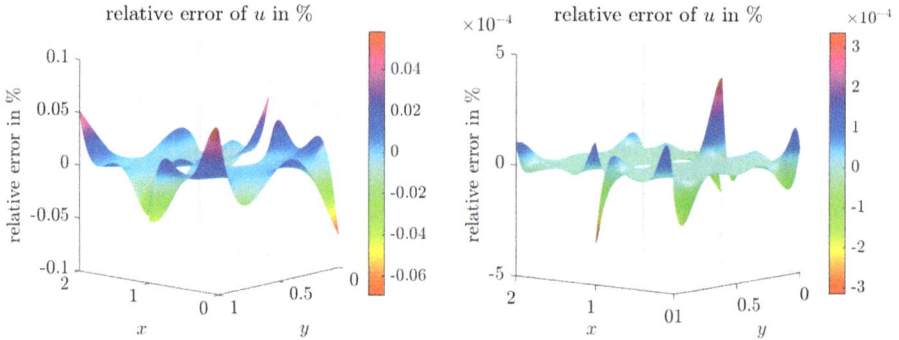

Figure 13: Relative error of the potential of the local expansion computed using the multipole-to-local transformation for $L = 5$ and $L = 10$.

4 CONCLUSIONS

In summary, my here presented studies on spherical harmonics and the fast multipole method deepen the understanding of both and therefore support more efficient use of these powerful tools. First, I have revisited spherical harmonics and particularly showed that spherical harmonics even for high degrees have coordinate dependencies in Cartesian coordinates although their input arguments are defined in spherical coordinates. Therefore, the coefficients of the related truncated series expansion enable accurate control of the intended approximation concerning dependencies in Cartesian coordinates. Then, I have discussed the reversed FMM algorithm in detail and theoretically showed its efficiency in the context of meshfree post-processing. Finally, I have demonstrated based on an example relevant to practice that with typical FMM parameters highly accurate computations are possible with the FMM. I could experimentally prove that the influence of transformations of multipole and local coefficients is negligible. The total accuracy of the FMM is mainly influenced by distances between the related octree cubes and the actual distance between boundary elements.

REFERENCES

[1] Greengard, L. & Rokhlin, V., The rapid evaluation of potential fields in three dimensions. *Vortex Methods*, Lecture Notes in Mathematics, vol. 1360, eds C. Anderson & C. Greengard, Springer, 1988. DOI: 10.1007/BFb0089775.

[2] Nabors, K. & White, J., FastCap: A multipole accelerated 3-D capacitance extraction program. *IEEE Transactions on Computer-Aided Design of Integrated Circuits and Systems*, **10**(11), pp. 1447–1459, 1991. DOI: 10.1109/43.97624.

[3]   Buchau, A., Rieger, W. & Rucker, W.M., BEM computations using the fast multipole method in combination with higher order elements and the Galerkin method. *IEEE Transactions on Magnetics*, **37**(5), pp. 3181–3185, 2001. DOI: 10.1109/20.952572.

[4]   Buchau, A., Rucker, W.M., Rain, O., Rischmüller, V., Kurz, S. & Rjasanow, S., Comparison between different approaches for fast and efficient 3-D BEM computations. *IEEE Transactions on Magnetics*, **39**(3), pp. 1107–1110, 2003. DOI: 10.1109/TMAG.2003.810167.

[5]   Brebbia, C.A., Telles, J.C.F. & Wrobel, L.C., *Boundary Element Techniques*, Springer-Verlag: Berlin and New York, 1984.

[6]   Buchau, A., Precise and robust magnetic field computations for high-end smart sensor applications. *WIT Transactions on Engineering Sciences*, vol. 126, WIT Press: Southampton and Boston, pp. 75–87, 2019. DOI: 10.2495/BE420071.

[7]   Buchau, A. & Rucker, W.M., Meshfree computation of field lines across multiple domains using fast boundary element methods. *IEEE Transactions on Magnetics*, **51**(3), pp. 1–4, 2015. DOI: 10.1109/TMAG.2014.2359520.

[8]   Greengard, L. & Rokhlin, V., A new version of the fast multipole method for the Laplace equation in three dimensions. *Acta Numerica*, **6**, pp. 229–269, 1997. DOI: 10.1017/S0962492900002725.

[9]   Arfken, G.B. & Weber, H.J., *Mathematical Methods for Physicists*, Academic Press, 1995.

[10] Hobson, E.W., *The Theory of Spherical and Ellipsoidal Harmonics*, Cambridge University Press, 1931.

[11] Maxwell, J.C., *Treatise on Electricity and Magnetism*, Clarendon Press, 1873.

[12] Buchau, A. & Rucker, W.M., Feasibility of a meshfree post-processing for boundary element methods. *WIT Transactions on Modelling and Simulation*, vol. 61, WIT Press: Southampton and Boston, pp. 327–338, 2015. DOI: 10.2495/BEM380261.

ENERGETIC GALERKIN BOUNDARY ELEMENT METHOD FOR 2D ELASTODYNAMICS: INTEGRAL OPERATORS WITH WEAK AND STRONG SINGULARITIES

GIULIA DI CREDICO, ALESSANDRA AIMI & CHIARA GUARDASONI
Department of Mathematical, Physical and Computer Sciences, University of Parma, Italy

ABSTRACT

In this paper, we consider some elastodynamics problems in 2D unbounded domains, with soft scattering conditions at the boundary, and their solution by the Boundary Element Method (BEM). The displacement identifying the elastic wave propagation is represented by both direct and indirect boundary integral formulations, which depend on the traction or on the jump of the traction at the boundary of the propagation domain, respectively. We study the characteristic singularities of the single layer and the double layer integral operators, which are involved in the considered *energetic* weak forms. Some algorithmic considerations about the parallel implementation of the *energetic* BEM and the quadrature techniques applied to overcome the issues due to the weak and the strong singularities of the integration kernels are proposed. Numerical simulations follow, showing a comparison between the external displacements obtained by the indirect and the direct formulations.

Keywords: elastodynamics, energetic BEM, single layer operator, double layer operator, weakly singular kernel, strongly singular kernel.

1 INTRODUCTION

The study of elastodynamics has several applications, including mechanical and civil engineering problems, seismic risk assessment and geological soil analyses. Often these problems are approached by domain methods, however this is not an optimal choice if the propagation happens throughout an external and unbounded domain, since artificial boundary conditions need to be imposed, potentially causing spurious reflections. The Boundary Element Method (BEM), instead, is in this context an efficient numerical technique, because both outer and interior propagations can be incorporated in the approximate solution of an integral equation, defined on the boundary of the diffusion domain. In this paper, we implement an energetic version of the BEM, introduced and analyzed for elastodynamics in the pioneering work by Bécache and Ha-Duong [1], in order to overcome the instabilities arising from the application of the standard BEMs [2] and with the aim of extending the optimal results for soft scattering problems reported by some authors of this work in [3].

2 MODEL PROBLEM

We study the propagation of an elastic vectorial wave in a two-dimensional domain $\Omega_e \subset \mathbb{R}^2$, considering both the following situations: Ω_e external to a bounded and simply connected domain Ω_i with Lipschitz boundary $\Gamma = \partial\Omega_i$, namely $\Omega_e = \mathbb{R}^2 \setminus \overline{\Omega}_i$, or Ω_e external to an open arc Γ, namely $\Omega_e = \mathbb{R}^2 \setminus \Gamma$. Under the assumption of null external body forces, the displacement field $\mathbf{u}(\mathbf{x}, t) = (u_1, u_2)^\top(\mathbf{x}, t)$, $\mathbf{x} = (x_1, x_2)^\top \in \Omega_e$, satisfies the *Navier equation* written in terms of displacement components [4]:

$$\sum_{h,k,l=1}^{2} \frac{\partial}{\partial x_h}\left(C_{ih}^{kl}\frac{\partial u_k}{\partial x_l}(\mathbf{x}, t)\right) - \varrho \ddot{u}_i(\mathbf{x}, t) = 0, \ \mathbf{x} \in \Omega_e, t \in (0, T], i = 1,2, \tag{1}$$

WIT Transactions on Engineering Sciences, Vol 131, © 2021 WIT Press
www.witpress.com, ISSN 1743-3533 (on-line)
doi:10.2495/BE440021

where ϱ is the mass density and the upper dots indicate derivatives with respect to time. The *Hooke tensor* $C_{ih}^{kl} = \lambda\delta_{ih}\delta_{kl} + \mu(\delta_{ik}\delta_{hl} + \delta_{il}\delta_{hk})$ depends on the *Kronecker Delta* δ_{ij} and on the positive *Lamé parameters* λ and μ, incorporating physical information about the diffusion domain. T is the final time instant of analysis. We further introduce the traction field $\mathbf{p}(\mathbf{x}, t) = (p_1, p_2)^\top(\mathbf{x}, t)$ defined on the obstacle Γ:

$$p_i(\mathbf{x}, t) = \sum_{h,k,l=1}^{2} C_{ih}^{kl}\frac{\partial u_k}{\partial x_l}(\mathbf{x}, t)n_{\mathbf{x}h}, \quad \mathbf{x} \in \Gamma, t \in (0, T], i = 1,2,$$

where $\mathbf{n_x} = (n_{\mathbf{x}1}, n_{\mathbf{x}2})^\top$ is the unit outward normal vector to Γ (if Γ is an open arc and if Γ^- and Γ^+ denote respectively the lower and the upper sides of the obstacle, then $\mathbf{n_x}$ is oriented from Γ^- to Γ^+). Eqn (1) is complemented by initial vanishing conditions and by Dirichlet boundary conditions, modelling a *soft scattering* by the obstacle, i.e.:

$$\mathbf{u}(\mathbf{x}, 0) = \dot{\mathbf{u}}(\mathbf{x}, 0) = 0, \qquad\qquad\qquad \mathbf{x} \in \Omega_e, \tag{2}$$

$$\mathbf{u}(\mathbf{x}, t) = \mathbf{g}(\mathbf{x}, t), \qquad (\mathbf{x}, t) \in \Sigma := \Gamma \times (0, T]. \tag{3}$$

3 REPRESENTATION FORMULA AND BOUNDARY INTEGRAL EQUATIONS

To describe the unknown $\mathbf{u}$ in $\Omega_e \times (0, T]$, we consider both direct and indirect *integral representation equations*:

$$\mathbf{u}(\mathbf{x}, t) = \mathcal{K}\mathbf{u}(\mathbf{x}, t) - \mathcal{V}\mathbf{p}(\mathbf{x}, t), \tag{4}$$

$$\mathbf{u}(\mathbf{x}, t) = \mathcal{V}\boldsymbol{\phi}(\mathbf{x}, t). \tag{5}$$

The direct eqn (4) depends on the behaviour of the traction field $\mathbf{p}$ on Γ, while in the indirect eqn (5) the unknown density vector $\boldsymbol{\phi} = (\phi_1, \phi_2)^\top$ corresponds to the jump of $\mathbf{p}$ along the arc Γ. Setting a Dirichlet condition as in eqn (3), the external problem can be solved using, alternatively, eqns (4) or (5). The *single layer operator* $\mathcal{V}$ and the *double layer operator* $\mathcal{K}$ act on a generic field $\boldsymbol{\psi}$ defined on Σ as follows:

$$[\mathcal{V}\boldsymbol{\psi}]_i(\mathbf{x}, t) = \sum_{j=1}^{2}\int_0^t\int_\Gamma G_{ij}(\mathbf{x}, \boldsymbol{\xi}; t, \tau)\psi_j(\boldsymbol{\xi}, \tau)\, d\Gamma_{\boldsymbol{\xi}}d\tau,$$

$$[\mathcal{K}\boldsymbol{\psi}]_i(\mathbf{x}, t) = \sum_{j=1}^{2}\int_0^t\int_\Gamma\sum_{h,k,l=1}^{2} C_{jh}^{kl}\frac{\partial G_{ij}}{\partial \xi_l}(\mathbf{x}, \boldsymbol{\xi}; t, \tau)\psi_j(\boldsymbol{\xi}, \tau)n_{h\boldsymbol{\xi}}\, d\Gamma_{\boldsymbol{\xi}}d\tau, \tag{6}$$

$$(\mathbf{x}, t) \in \Omega_e \times (0, T], \ i = 1,2.$$

The second order tensor G_{ij} in eqn (6) is the *Green tensor*, that is, the fundamental solution of the considered differential problem:

$$G_{ij}(\mathbf{x}, \boldsymbol{\xi}; t, \tau) := \frac{H[c_P(t-\tau)-r]}{2\pi\varrho c_P}\left\{\frac{r_ir_j}{r^4}\frac{2c_P^2(t-\tau)^2-r^2}{\sqrt{c_P^2(t-\tau)^2-r^2}} - \frac{\delta_{ij}}{r^2}\sqrt{c_P^2(t-\tau)^2-r^2}\right\}$$

$$-\frac{H[c_S(t-\tau)-r]}{2\pi\varrho c_S}\left\{\frac{r_ir_j}{r^4}\frac{2c_S^2(t-\tau)^2-r^2}{\sqrt{c_S^2(t-\tau)^2-r^2}} - \frac{\delta_{ij}}{r^2}\frac{c_S^2(t-\tau)^2}{\sqrt{c_S^2(t-\tau)^2-r^2}}\right\}, \tag{7}$$

where $H[\cdot]$ is the *Heaviside function* and where $c_P = \sqrt{(\lambda + 2\mu)/\varrho} > 0$ and $c_S = \sqrt{\mu/\varrho} > 0$ are the fundamental velocities of the irrotational and rotational displacements in which the vectorial solution of eqn (1) can be decomposed. The Green tensor G_{ij} is symmetric w.r.t the space variables since it depends on them only through the vector $\mathbf{r} = (r_1, r_2)^\top = \mathbf{x} - \boldsymbol{\xi} = (x_1 - \xi_1, x_2 - \xi_2)^\top$ and on its norm $r = \|\mathbf{r}\|_2$, namely the distance between the field $\mathbf{x}$ and the source point $\boldsymbol{\xi}$. Considering a point $\mathbf{x} \in \Gamma$, from the representation eqns (4) and (5), we obtain the following *Boundary Integral Equations* (BIEs):

$$\mathcal{V}\mathbf{p}(\mathbf{x},t) = \left(\mathcal{K} - \tfrac{1}{2}\mathcal{J}\right)\mathbf{g}(\mathbf{x},t), \quad (\mathbf{x},t) \in \Sigma, \tag{8}$$

$$\mathcal{V}\boldsymbol{\phi}(\mathbf{x},t) = \mathbf{g}(\mathbf{x},t), \quad (\mathbf{x},t) \in \Sigma, \tag{9}$$

where $\mathcal{J}$ is the identity operator and $\mathbf{g} = (g_1, g_2)^\top$ is the boundary datum assigned in eqn (3).

4 ENERGETIC WEAK FORMULATION AND SPACE-TIME DISCRETIZATION
In this work we reformulate the integral eqns (8) and (9) in the so called *energetic* weak form: this approach, already analyzed for scalar wave propagation problems [5] and introduced for soft scattering elastodynamic problems in [3], allows one to overcome the instabilities arising from the standard weak form of the BIEs. Therefore, having set $W = \left(L^2\left([0,T]; H^{-1/2}(\Gamma)\right)\right)^2$, the weak problems we intend to solve are as follows: *find $\boldsymbol{\varphi} \in W$ such that*

$$\langle \mathcal{V}\boldsymbol{\varphi}, \dot{\boldsymbol{\psi}} \rangle_{L^2(\Sigma)} = \langle \bar{\mathbf{g}}, \dot{\boldsymbol{\psi}} \rangle_{L^2(\Sigma)}, \quad \forall \boldsymbol{\psi} \in W, \tag{10}$$

where $\boldsymbol{\varphi}$ coincides with $\mathbf{p}$ or $\boldsymbol{\phi}$ and $\bar{\mathbf{g}}$ coincides with $(\mathcal{K} - 1/2\,\mathcal{J})\,\mathbf{g}$ or $\mathbf{g}$ in case we are solving eqn (8) or (9) respectively.

To solve the weak problem (10) in a discretized form, we first consider a uniform decomposition of the time interval $[0,T]$, with time step $\Delta t = T/N_{\Delta t}$, $N_{\Delta t} \in \mathbb{N}^+$ and setting $N_{\Delta t} + 1$ time-knots $t_n = n\Delta t$, $n = 0, \ldots, N_{\Delta t}$. Having defined $\mathcal{P}_d$ as the space of algebraic polynomials of degree d, we define the corresponding space of piecewise polynomial functions of degree d_t in time

$$\mathcal{T}_{\Delta t, d_t} = \left\{\sigma : \sigma|_{[t_k, t_{k+1}]} \in \mathcal{P}_{d_t} \; \forall_{k=0}^{N_{\Delta t}-1}, \sigma \in C^0([0,T]) \text{ and } \sigma(0) = 0 \text{ if } d_t \geq 1\right\}.$$

For the space discretization we introduce a boundary mesh constituted by a set of straight consecutive elements $\mathcal{E} = \{e_1, \ldots, e_M\}$ with length $2l_i := length(e_i) \leq \Delta x$, $e_i \cap e_j = \emptyset$ if $\neq j$. For a piecewise linear arc, the union of the mesh elements is equivalent to Γ, or it represents a suitable approximation of the boundary in other cases. We have to choose functions in $H^{-1/2}(\Gamma)$ for the spatial approximation, therefore we consider the space of piecewise polynomial functions

$$X_{\Delta x, d_x} = \left\{w \in L^2(\Gamma) : w|_{e_i} \in \mathcal{P}_{d_x}, e_i \in \mathcal{E}\right\}.$$

The Galerkin approximation of the energetic weak forms (10) in these spaces, having set $W_{\Delta x, \Delta t} = \left(X_{\Delta x, d_x} \otimes \mathcal{T}_{\Delta t, d_t}\right)^2$, reads:

Find $\boldsymbol{\varphi}_{\Delta x,\Delta t} \in W_{\Delta x,\Delta t}$ such that

$$\langle \mathcal{V}\, \boldsymbol{\varphi}_{\Delta x,\Delta t}, \dot{\boldsymbol{\psi}}_{\Delta x,\Delta t}\rangle_{L^2(\Sigma)} = \langle \overline{\mathbf{g}}, \dot{\boldsymbol{\psi}}_{\Delta x,\Delta t}\rangle_{L^2(\Sigma)}, \; \forall\, \dot{\boldsymbol{\psi}}_{\Delta x,\Delta t} \in W_{\Delta x,\Delta t}. \tag{11}$$

We introduce the set $\{w_m(\mathbf{x})\}_{m=1}^{M_{\Delta x}}$, containing the basis functions of the space $X_{\Delta x,d_x}$, which are piecewise polynomials depending on the Lagrangian polynomials on each element e_i. For the time discretization we choose piecewise constant basis functions ($d_t = 0$):

$$\chi_n(t) = H[t - t_n] - H[t - t_{n+1}], \quad n = 0, \dots, N_{\Delta t} - 1.$$

Hence, the components of the discrete function $\boldsymbol{\varphi}_{\Delta x,\Delta t}$ can be expressed in space and time as

$$\varphi_{\Delta x,\Delta t}^i(\mathbf{x}, t) = \sum_{n=0}^{N_{\Delta t}-1} \sum_{m=1}^{M_{\Delta x}} \alpha_{nm}^i w_m(\mathbf{x})\chi_n(t), \quad i = 1,2.$$

Replacing the test functions with the space-time basis functions defined above, the Galerkin eqn (11) leads to the linear system $E\boldsymbol{\alpha} = \boldsymbol{\beta}$, with the following block structure

$$\begin{pmatrix} E^{(0)} & 0 & \cdots & 0 \\ E^{(1)} & E^{(0)} & \cdots & 0 \\ \vdots & \vdots & \ddots & \vdots \\ E^{(N_{\Delta t}-1)} & E^{(N_{\Delta t}-2)} & \cdots & E^{(0)} \end{pmatrix} \begin{pmatrix} \boldsymbol{\alpha}_0 \\ \boldsymbol{\alpha}_1 \\ \vdots \\ \boldsymbol{\alpha}_{N_{\Delta t}-1} \end{pmatrix} = \begin{pmatrix} \boldsymbol{\beta}_0 \\ \boldsymbol{\beta}_1 \\ \vdots \\ \boldsymbol{\beta}_{N_{\Delta t}-1} \end{pmatrix}, \tag{12}$$

where for all $l = 0, \dots, N_{\Delta t} - 1$ the l-th block and the l-th entry of the solution vector are organized as

$$E^{(l)} = \begin{pmatrix} E_{11}^{(l)} & E_{12}^{(l)} \\ E_{21}^{(l)} & E_{22}^{(l)} \end{pmatrix}, \quad \boldsymbol{\alpha}^{(l)} = \begin{pmatrix} \alpha_{l1}^1 & \cdots & \alpha_{lM_{\Delta x}}^1 & \alpha_{l1}^2 & \cdots & \alpha_{lM_{\Delta x}}^2 \end{pmatrix}.$$

The structure of the right-hand side $\boldsymbol{\beta}$ is analogous to that of the solution vector.
Solving eqn (12) by blockwise forward substitution leads to a marching-on-in-time (MOT) time stepping scheme. The double analytical integration in the time variables, leads to the matrix entry

$$\left(E_{ij}^{(l)}\right)_{\widetilde{m},m} = -\sum_{\zeta,\varsigma=0}^{1} \frac{(-1)^{\zeta+\varsigma}}{2\pi\varrho} \int_\Gamma \int_\Gamma w_{\widetilde{m}}(\mathbf{x}) w_m(\boldsymbol{\xi}) v_{ij}\left(r; \Delta_{\tilde{n}+\zeta,n+\varsigma}\right) d\Gamma_{\boldsymbol{\xi}} d\Gamma_{\mathbf{x}}, \tag{13}$$

for all $i,j = 1,2$, $m, \widetilde{m} = 1, \dots, M_{\Delta x}$ and $n, \tilde{n} = 0, \dots, N_{\Delta t} - 1$, where the positive time difference $t_{\tilde{n}+\zeta} - t_{n+\varsigma} = \Delta_{\tilde{n}+\zeta,n+\varsigma}$, with $\tilde{n} - n = l$, is set. For a generic $\Delta > 0$, the kernel v_{ij} is defined as

$$v_{ij}(r; \Delta) := \left(\frac{r_i r_j}{r^4} - \frac{\delta_{ij}}{2r^2}\right)\left[\frac{H[c_P\Delta - r]}{c_P} \Delta h_{P,1}(r; \Delta) - \frac{H[c_S\Delta - r]}{c_S} \Delta h_{S,1}(r; \Delta)\right]$$
$$+ \frac{\delta_{ij}}{2}\left[\frac{H[c_P\Delta - r]}{c_P^2} h_{P,2}(r; \Delta) + \frac{H[c_S\Delta - r]}{c_S^2} h_{S,2}(r; \Delta)\right], \tag{14}$$

with the specific kernel functions given by

$$h_{\gamma,1}(r;\Delta) := \sqrt{c_\gamma^2\Delta^2 - r^2}, \quad h_{\gamma,2}(r;\Delta) := \log\left(\sqrt{c_\gamma^2\Delta^2 - r^2} + c_\gamma\Delta\right) - \log(r), \quad \gamma = S, P.$$

If $0 \le r \le c_S\Delta < c_P\Delta$ all the arguments of the Heaviside functions in eqn (14) are positive and the kernel v_{ij} can be expressed in a reduced form with space singularity of kind $\mathcal{O}(\log(r))$ for $r \to 0$. This behaviour is well-studied for boundary integral operators related to 2D elliptic problems. If we choose to solve the discretized weak form arising from eqn (9), the right hand side $\boldsymbol{\beta}$ has elements made by the $L^2(\Sigma)$ product of the Dirichlet datum $\mathbf{g}$ by the time derivative of the basic functions in space and time. Otherwise, starting from eqn (8), the product $\langle \mathcal{K}\mathbf{g}, \dot{\boldsymbol{\psi}}\rangle_{L^2(\Sigma)}$ is evaluated by an approximation of the two components of the boundary datum with a linear combination of space–time basis functions:

$$g_i(\mathbf{x}, t) \approx \hat{g}_i(\mathbf{x}, t) := \sum_{n=0}^{N_{\Delta t}-1} \sum_{m=1}^{M_{\Delta x}} \hat{g}_{nm}^i w_m(\mathbf{x}) v_n(t), \qquad i = 1,2,$$

where

$$v_n(t) = H[t - t_n]\frac{t - t_n}{\Delta t} - H[t - t_{n+1}]\frac{t - t_{n+1}}{\Delta t}, \quad n = 0, \dots, N_{\Delta t} - 1,$$

are basis functions of $\mathcal{T}_{\Delta t,1}$. To have a correct interpolation of the boundary datum, the condition $\hat{g}_i(\mathbf{x}_m, t_k) = g_i(\mathbf{x}_m, t_k)$ is imposed: the spatial nodes $\mathbf{x}_m$ are the middle points of the mesh elements e_m if w_m are piecewise constant, otherwise are the amount of points determined by the interpolation nodes of the Lagrangian polynomials defining spatial shape functions of higher degree. With this substitution, the evaluation of $\langle \mathcal{K}\mathbf{g}, \dot{\boldsymbol{\psi}}\rangle_{L^2(\Sigma)}$ becomes the computation of the matrix-vector product $\hat{E}\hat{\mathbf{g}}$, with $\hat{E}$ having similar Toeplitz structure as E and $\hat{\mathbf{g}}$ column vector with elements $\hat{g}_{nm}^i$. Also for the double layer operator, we can perform an analytical time integration, leading to the matrix entry

$$\left(\hat{E}_{ij}^{(l)}\right)_{\tilde{m},m} = \sum_{\zeta,\varsigma=0}^{1} \frac{(-1)^{\zeta+\varsigma}}{\pi\varrho\Delta t} \int_\Gamma \int_\Gamma w_{\tilde{m}}(\mathbf{x}) w_m(\boldsymbol{\xi}) \hat{v}_{ij}\left(r; \Delta_{\tilde{n}+\zeta, n+\varsigma}\right) d\Gamma_{\boldsymbol{\xi}} d\Gamma_{\mathbf{x}}, \tag{15}$$

for all $i, j = 1,2, m, \tilde{m} = 1, \dots, M_{\Delta x}$ and $n, \tilde{n} = 0, \dots, N_{\Delta t} - 1$. The integration kernel $\hat{v}_{ij}$, for a positive Δ, is defined as follows

$$\hat{v}_{ij}(r;\Delta) := \sum_{h,k,l=1}^{2} C_{jh}^{kl}\left(-\frac{A_{ikl}}{2r}\frac{h_{S,1}(r;\Delta)}{c_S^3} H[c_S\Delta - r]\right.$$
$$+ \frac{1}{6r} B_{ikl}\left[\frac{h_{S,1}(r;\Delta)}{c_S^3} H[c_S\Delta - r] - \frac{h_{P,1}(r;\Delta)}{c_P^3} H[c_P\Delta - r]\right]$$
$$+ \left.\frac{\Delta^2}{6r^3} D_{ikl}\left[\frac{h_{P,1}(r;\Delta)}{c_P} H[c_P\Delta - r] - \frac{h_{S,1}(r;\Delta)}{c_S} H[c_S\Delta - r]\right]\right) n_{h\xi},$$

with $A_{ikl} = \delta_{ik}r_{,l}$, $B_{ikl} = \delta_{il}r_{,k} + \delta_{kl}r_{,i} + \delta_{ik}r_{,l} - r_{,i}r_{,k}r_{,l}$, $D_{ikl} = \delta_{il}r_{,k} + \delta_{kl}r_{,i} + \delta_{ik}r_{,l} - 4r_{,i}r_{,k}r_{,l}$ and $r_{,k} = r_k/r$ for $k = 1,2$. If $0 \le r \le c_S\Delta < c_P\Delta$, the kernel $\hat{v}_{ij}$ reduces to a simplified form with strong singularity $\mathcal{O}(1/r)$ for $r \to 0$.

5 PARALLEL COMPUTATION OF TOEPLITZ MATRICES

The construction of the linear system (12) requires just the computation of the first column of blocks of E and this is a great advantage in terms of computations saving and memory requirement. Moreover, the computation of the matrix blocks $E^{(l)}$ and of the right hand side $\boldsymbol{\beta}_l$ is embarrassingly parallel. For this reason, the entire method can be improved in programming phase with a parallel computing approach. The elapsed times decay reported in Fig. 1 refers to the construction of 400 time blocks of E and $\boldsymbol{\beta}$, each depending on 320 spatial degrees of freedom. Increasing the number of threads used to compute every block of the linear system leads to a linear decline of the execution times.

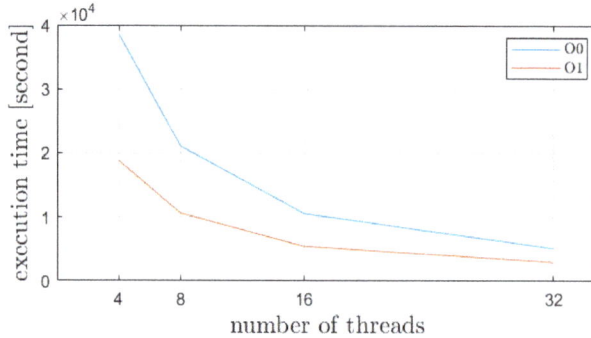

Figure 1:   Scalability of the algorithm that computes the time blocks of the matrix and of the right hand side of the linear system $E\boldsymbol{\alpha} = \boldsymbol{\beta}$ (in eqn (11) $\bar{\mathbf{g}} = \mathbf{g}$ is set).

In particular the method is implemented by the parallelization of a Fortran code with open-mp library. The labels $O0, O1$ in Fig. 1 refer to different optimization setup of the Intel compiler: the options $O0$ is the default, the option $O1$ maximizes speed by negligibly reducing the accuracy, leading to a faster parallel computation of the time block matrices. If the problem is solved with a direct approach, similar considerations can be repeated for the parallel computation of the double layer matrix $\hat{E}$.

6 INTEGRATION TECHNIQUES

Following the element by element technique, we calculate the double integral in eqn (13), for a positive time difference Δ, as the sum of local contributions of type

$$\int_0^{2l_{\widetilde{m}}} \int_0^{2l_m} w_{\widetilde{m}}^{(d_x)}(s) w_m^{(d_x)}(z) v_{ij}(r(s,z);\Delta) dz ds, \tag{16}$$

obtained after the parametrization of the generic integration segments e_m and $e_{\widetilde{m}}$ belonging to Γ discretization (see Section 4):

$$\mathbf{x} \to \ s \in [0,2l_m], \qquad \boldsymbol{\xi} \to z \in [0,2l_{\widetilde{m}}],$$

and where $w_{\widetilde{m}}^{(d_x)}(s)$ and $w_m^{(d_x)}(z)$ are local Lagrangian basis functions in the space variable defined over the elements $e_{\widetilde{m}}$ and e_m. The double integral in eqn (15) can be written in the same way by substituting v_{ij} with $\hat{v}_{ij}$. An important integration issue is the dependence of both kernels v_{ij} and $\hat{v}_{ij}$ on the Heaviside functions $H[c_\gamma\Delta - r]$, which define the wavefronts with peculiar velocities c_S and c_P. To avoid the integration of these discontinuous functions, the local rectangular domain of integration $[0,2l_{\tilde{m}}] \times [0,2l_m]$ in eqn (16) is further divided into regions depending on the splitting points determined by the equation $r = c_\gamma \Delta$, $\gamma = S, P$, where the distance r can assume different expressions:

$$e_{\tilde{m}} \equiv e_m \Longrightarrow r = |s - z|,$$
$$e_{\tilde{m}}, e_m \text{ consecutive and aligned } (m < \tilde{m}) \Longrightarrow r = s + 2l_m - z,$$
$$e_{\tilde{m}}, e_m \text{ consecutive and not aligned with angle } \theta \ (m < \tilde{m})$$
$$\Longrightarrow r^2 = s^2 + (2l_m - z)^2 - 2\cos(\theta)\,s(2l_m - z).$$

An example of integration domains depending on the mutual position of the mesh elements is in Fig. 2, where in particular the sets $E_S \cap [0,2l_{\tilde{m}}] \times [0,2l_m]$ and $E_{S,P} \cap [0,2l_{\tilde{m}}] \times [0,2l_m]$, with

$$E_S = \{r(s,z) < c_S\Delta\}, \qquad E_{S,P} = \{c_S\Delta < r(s,z) < c_P\Delta\},$$

are highlighted.

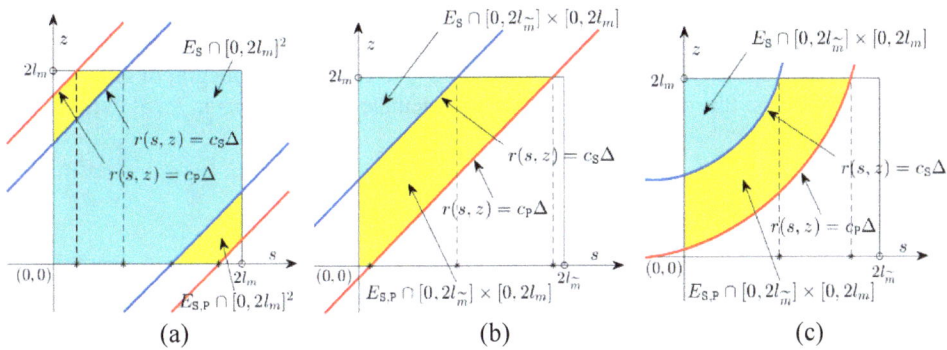

Figure 2: Example of integration domains in case of coincident elements (a), consecutive aligned elements (b) and consecutive not aligned elements (c).

Then, we calculate eqn (16) summing the contribution over each region, where s belongs to a specific interval $[a, b] \subset [0,2l_{\tilde{m}}]$, and the inner interval $[A(s), B(s)]$ strictly depends on the outer variable. A complete overview of the shape and the related splitting of the integration domains is reported in a recently submitted work [6].

We focus now on the single layer kernel v_{ij}, observing that it depends on the function $\log\big(c_\gamma\Delta + \sqrt{c_\gamma^2\Delta^2 - r^2}\big), \gamma = S, P$, which is not singular but has an unbounded derivative for $r = c_\gamma\Delta$. Therefore, the accurate integration of this part of the kernel would require a really large number of Gauss-Legendre quadrature nodes. To avoid high computational costs, the inner integration of this function is computed by combining the Gaussian rule with a regularization procedure, analyzed in [7], involving the use of the following change of integration variable $= \Phi(\hat{z})$, $\hat{z} \in [0,1]$, where

$$\Phi(\hat{z}) = \frac{(p+q-1)!}{(p-1)!(q-1)!} \int_0^{\hat{z}} u^{p-1}(1-u)^{q-1}du \tag{17}$$

helps to move the quadrature nodes towards the endpoints of the interval (where it is possible to have $r = c_\gamma \Delta$), in order to approximate better the behaviour of the square root. The position of the nodes is regulated by the value of the characteristic exponents $p, q \geq 1$ of the equation. The outer integration in the s variable can be performed with a standard Gauss-Legendre rule.

Further integration problems arise from logarithmic singularity. In case of coincident elements, where $r = |s - z|$ is null for $s = z$, it is possible to compute the inner integration with the quadrature equations proposed in [8], of the type

$$\int_{-1}^{1} log(|s - z_G|) f(z_G) dz_G \approx \sum_{i=1}^{n} \omega_i(s) f(z_{G,k}). \tag{18}$$

The weights in eqn (18) depend both on the weights of the Gauss–Legendre equation and on the singular kernel, while the nodes $z_{G,k}$ are the roots of the Legendre polynomials of degree n. The recursive relation to calculate $\omega_i(s)$ may be found in [8]. The equation is interpolatory, namely it calculates the integral exactly if the function f is a polynomial up to degree $n - 1$, making the procedure suitable for our purposes since the kernel in eqn (16) is multiplied by the basic function $w_m^{(dx)}$. Moreover, eqn (18) can be easily adapted to other mutual positions of the mesh elements.

Concerning the double layer kernel $\hat{v}_{ij}$, we have to deal with the strong singularity $\mathcal{O}(1/r)$. Let us suppose also in this case that we have to compute an integral of type (16) for coincident elements: this means that we have to calculate contributions in local variables of the following type

$$I = \int_a^b \int_{A(S)}^{B(S)} w_{\tilde{m}}^{(dx)}(s) w_m^{(dx)}(z) \frac{\sqrt{c_S^2 \Delta^2 - (s-z)^2}}{s-z} dz ds, \tag{19}$$

where in particular the kernel is strongly singular at all points (s, z) in the integration domain where $z = s$. If we operate a first-order Taylor expansion of the numerator of the interior integrand function for $z = s$, integral (19) becomes

$$I = -\int_a^b w_{\tilde{m}}^{(dx)}(s) \int_{A(S)}^{B(S)} \frac{w_m^{(dx)}(z) \sqrt{c_S^2 \Delta^2 - (s-z)^2} - c_S \Delta w_m^{(dx)}(s)}{z-s} dz ds$$
$$-\int_a^b w_{\tilde{m}}^{(dx)}(s) c_S \Delta w_m^{(dx)}(s) \int_{A(S)}^{B(S)} \frac{1}{z-s} dz ds = I_1 + I_2. \tag{20}$$

Integral I_1 in eqn (20) is no more singular but it requires the regularization procedure (17) if the argument of $\sqrt{c_S^2 \Delta^2 - (s - z)^2}$ vanishes at the endpoints of the integration domain. Regarding the second term I_2, the inner integral can be computed exactly:

$$\int_{A(S)}^{B(S)} \frac{1}{z-s} dz = log(B(s) - s) - log(s - A(s)). \tag{21}$$

The outer integration of I_2 presents an issue if the argument of the log functions in eqn (21) is null in one of the extremes of the interval $[a, b]$ (or in both). In case of these log mild singularities, we can perform the outer numerical integration combining a Gaussian quadrature rule with eqn (17), in order to suitably accumulate quadrature nodes towards the endpoints of integration.

7 NUMERICAL RESULTS

In this section we present two experiments which allow us to compare the results given by the discretization of the two formulations resumed in eqn (10). Concerning the parameters employed in all the following tests, we set peculiar velocities $c_S = 1$ and $c_P = 2$ and mass density $\varrho = 1$. The basis functions $\{w_m(\mathbf{x})\}_{m=1}^{M_{\Delta x}}$ for the space discretization are set to be piecewise constant (degree $d_x = 0$), meaning that the number $M_{\Delta x}$ of spatial degrees of freedom is equivalent to the number of elements e_m of the mesh over Γ. We recall that in this section we denote by $\mathbf{p}$ and $\boldsymbol{\phi}$ the approximations given by the solution of the energetic problem (11) related to the direct or indirect formulation respectively.

7.1 Experiment 1

In this example we consider a flat crack $\Gamma = \{(x_1, 0) \in \mathbb{R}^2 \mid -0.5 \le x_1 \le 0.5\}$ and the Dirichlet condition $\mathbf{g}(x_1, t)$ on Σ such that $g_1 \equiv 0$ and $g_2(x_1, t) = f(t + 0.45)$, where the temporal profile $f(t)$ is defined as follows:

$$f(t) = \begin{cases} t - 1/2, & t \in [0.5, 0.6] \\ (-30t + 19)/10, & t \in [0.6, 0.7] \\ (40t - 30)/10, & t \in [0.7, 0.8] \\ (-30t + 26)/10, & t \in [0.8, 0.9] \\ t - 1, & t \in [0.9, 1] \\ 0 & \text{otherwise} \end{cases}.$$

We set the final instant $T = 4$ and we impose on Γ a uniform decomposition with step $\Delta x = 0.025$ (40 segments) and a time step of size $\Delta t = 0.0125$. The boundary datum models a uniform scattering by the obstacle with vertical direction. In Fig. 3(a) and 3(b) the solution arrays $\mathbf{p}$ and $\boldsymbol{\phi}$ of the energetic weak form (11), considering the indirect and the direct formulations respectively, are shown: in particular there is the time history of the horizontal and the vertical components of $\mathbf{p}$ and $\boldsymbol{\phi}$ at the point $(-1/4, 0) \in \Gamma$. It is possible to observe that ϕ_1 is trivial. This is due to the fact that the condition $\mathbf{g}$ is completely vertical and, on a flat horizontal obstacle, the Green tensor defined in eqn (7) has null components for $i \ne j$.

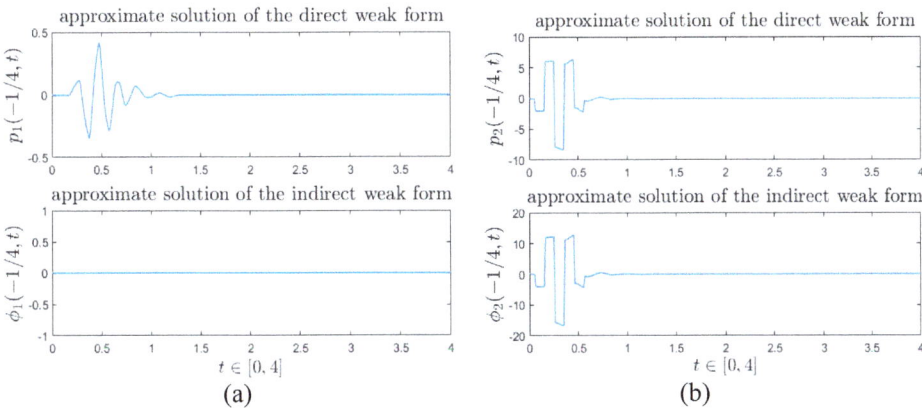

Figure 3: Horizontal and vertical components of the vector solutions $\mathbf{p}$ and $\boldsymbol{\phi}$ of the weak problem (11) calculated at point $(-1/4) \in \Gamma$ and during the time interval $[0,4]$.

This leads to a reduction of the indirect BIE (9), which is decoupled in two scalar equations: $\mathcal{V}_{11}\phi_1 = 0$ and $\mathcal{V}_{22}\phi_2 = g_2$. The right hand side of the direct weak form instead depends also on the double layer operator, meaning that the components of $\bar{\mathbf{g}}$ in eqn (11) are both active. Anyway, both approaches let us to calculate approximate results showing long time stability.

Solutions of the energetic weak formulations have then been employed in the respective integral representation eqns (4) and (5) to obtain the displacement $\mathbf{u}$ at the points of an axis orthogonal to Γ in its middle point $(0,0)$. Snapshots of the vertical displacement at the time instants 1, 2 and 3 are shown in Fig. 4: in Fig. 4(a) the component u_2 is calculated replacing the approximate traction $\mathbf{p}$ in the direct eqn (4), while Fig. 4(b) have been obtained by the indirect eqn (5), leading to a comparable results for the external propagation.

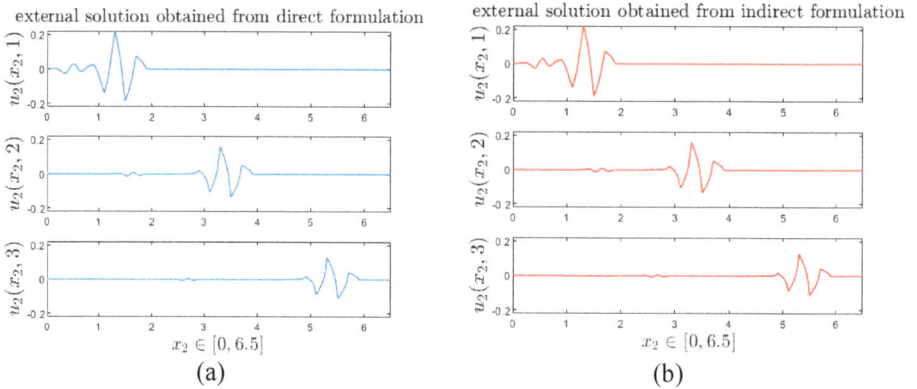

Figure 4:   External displacement $\mathbf{u}$ calculated, for different time instants, along the vertical axes $(0, x_2)$, $x_2 \in [0,6.5]$. Solutions have been obtained by the discretization of the direct weak form (a) and the indirect weak form (b).

This results in a wave spreading with the peculiar velocity $c_P = 2$, followed by a slower perturbation with speed $c_S = 1$. This last one is a circular perturbation having origin in the endpoints of the open arc Γ, usually called in literature crack tips phenomena.

Lastly, a numerical comparison is reported in Table 1, where the numbers represent the difference in L^2 norm between the external displacements obtained by the two different formulations evaluated at different time instants. These errors are $\mathcal{O}(10^{-3})$, as expected considering the magnitude of the imposed spatial and temporal discretization steps Δx and Δt.

Table 1: L^2-error for $x_2 \in [0,6.5]$ of the vertical components u_2 for some time instants.

$\left\|u_{2,direct}(\cdot,t) - u_{2,indirect}(\cdot,t)\right\|_{L^2([0,6.5])}$	$t = 1$	$t = 2$	$t = 3$
	$6.92 \cdot 10^{-3}$	$8.15 \cdot 10^{-4}$	$6.64 \cdot 10^{-4}$

7.2 Experiment 2

We now consider the circular arc $\Gamma = \{(x_1, x_2) \in \mathbb{R}^2 \mid x_1^2 + x_2^2 = 1\}$ and the Dirichlet condition $\mathbf{g}(\mathbf{x}, t)$ on Σ such that $g_1(\mathbf{x}, t) = h(t)x_1$ and $g_2(\mathbf{x}, t) = h(t)x_2$, where the function $h(t) = 0.2 \sin^2(40t) H[\pi/40 - t]$ is the temporal profile.

We set the final instant $T = 4$ and we impose on Γ a uniform decomposition with step $\Delta x \simeq 0.02$ (320 segments) and we choose a time step of size $\Delta t = 0.01$. With this problem geometry, the boundary datum models a uniform scattering by the obstacle with radial direction with respect to the centre of Γ. In Fig. 5 there is the time history of the vertical components of $\mathbf{p}$ and $\boldsymbol{\phi}$, obtained respectively by the discretized direct and indirect weak form (11), at the point $(0,1) \in \Gamma$. As we are solving an external problem, the traction $\mathbf{p}$ on Γ presents just an initial perturbation of length $\pi/40$, effect of the pulse generated by the Dirichlet boundary condition (see Fig. 5(a)). Differently, the density $\boldsymbol{\phi}$ in Fig. 5(b) presents a series of pulsations since it depends also on the interior wave reflections against Γ. In Fig. 6, three snapshots of the vertical external displacement $\mathbf{u}$ are represented, obtained respectively by the discretized direct and indirect weak form (11) (Fig. 6(a) and 6(b)), along the vertical axes $(0, x_2)$, $x_2 \in [1,8]$. Externally we have the diffusion of a pure radial perturbation with the peculiar velocity $c_P = 2$. There are no crack tips perturbations since the arc is closed.

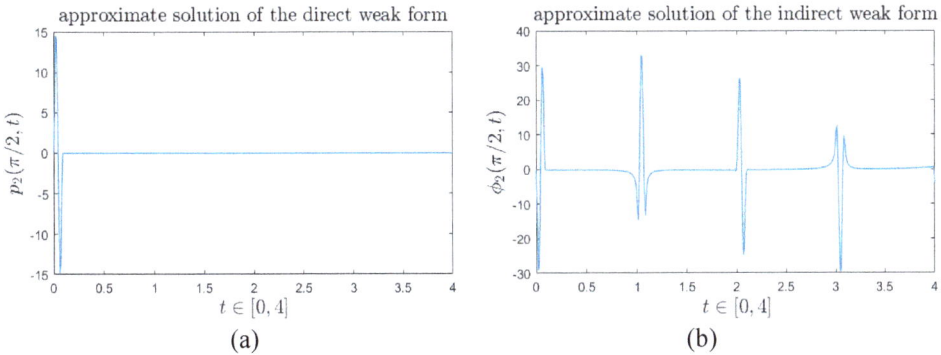

Figure 5: Vertical components of the vector solutions $\mathbf{p}$ and $\boldsymbol{\phi}$ of the weak problem (11) calculated at point $(0,1) \in \Gamma$ (corresponding to the clockwise angle $\pi/2$) and during the time interval $[0,4]$.

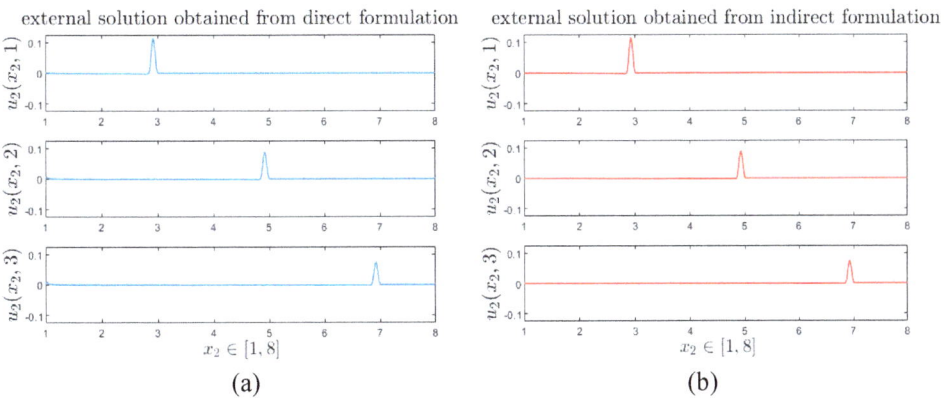

Figure 6: Vertical displacement u_2 calculated, for different time instants, along the vertical axes $(0, x_2)$, $x_2 \in [1,8]$. Solutions have been obtained by the discretization of the direct weak form (a) and the indirect weak form (b).

Also for the circular obstacle, we have a numerical comparison in Table 2, where the difference in L^2-norm between the external displacements obtained by the two different formulations is reported. Also in this case the errors are consistent with the fixed discretization space-time steps.

Table 2: L^2-error for $x_2 \in [1,8]$ of the vertical components u_2 for some time instants.

$\left\| u_{2,direct}(\cdot,t) - u_{2,indirect}(\cdot,t) \right\|_{L^2([1,8])}$	$t = 1$	$t = 2$	$t = 3$
	$5.32 \cdot 10^{-4}$	$1.36 \cdot 10^{-3}$	$2.07 \cdot 10^{-3}$

8 CONCLUSIONS

The comparison between the energetic BEM approximations of representation eqns (4) and (5) for the elastodynamic displacement, with soft scattering conditions, is the focus of this research.

By energetic BEM, the numerical results shown in Section 7 (concerning wave propagation problems external to a straight crack and to a closed circular domain) are achieved directly in the space-time domain allowing a better comprehension of the behaviour at the boundary of the solutions of the related direct and indirect BIEs (8) and (9).

The long-time stability of the results is independent of the discretization parameters. As described in Section 6, the efficiency and the accuracy of the entire energetic method can be equivalently maintained in both formulations by using suitable integration routines to treat the spatial singularities involved in the single and the double layer integration kernels and by the study of a proper splitting of the integration domains, determined by the two wave fronts characteristic of the elastic propagation. The direct eqn (4), even if more costly, will allow us to also handle the extension to mixed boundary condition problems.

As shown, the linear systems outcome of the energetic BEM applied to elastodynamic problems have a Toeplitz structure that leads to a considerable speed up during the execution, especially with parallel computation. From the point of view of the calculation time, this makes it easier to manage the increase of linear system degrees of freedom due to refinements of the space-time discretization steps.

ACKNOWLEDGEMENT

This work has been supported by INdAM, Italy, through granted GNCS research projects.

REFERENCES

[1] Bécache, E. & Ha Duong, T., A space-time variational formulation for the boundary integral equation in a 2d elastic crack problem. *ESAIM: Mathematical Modelling and Numerical Analysis*, **28**(2), pp. 141–176, 1994.

[2] Frangi, A. & Novati, G., On the numerical stability of time-domain elastodynamic analyses by BEM. *Computer Methods in Applied Mechanics and Engineering*, **173**(3–4), pp. 403–417, 1999.

[3] Aimi, A., Desiderio, L., Diligenti, M. & Guardasoni, C., Application of energetic BEM to 2D elastodynamic soft scattering problems. *Communications in Applied and Industrial Mathematics*, **10**(1), pp. 182–198, 2019.

[4] Andersen, L., *Linear Elastodynamic Analysis*, Lecture Notes, Department of Civil Engineering, Aalborg University, 2006.

[5] Aimi, A., Diligenti, M., Guardasoni, C., Mazzieri I. & Panizzi, S., An energy approach to space-time Galerkin BEM for wave propagation problems. *International Journal for Numerical Methods in Engineering*, **80**(9), pp. 1196–1240, 2009.

[6] Aimi, A., Di Credico, G., Diligenti, M. & Guardasoni, C., Highly accurate quadrature schemes for singular integrals in energetic BEM applied to elastodynamics. *Journal of Computational and Applied Mathematics* (submitted).

[7] Monegato, G. & Scuderi, L., Numerical integration of functions with boundary singularities. *Journal of Computational and Applied Mathematics*, **112**(1–2), pp. 201–214, 1999.

[8] Aimi A., Diligenti M., Monegato G., New numerical integration schemes for applications of Galerkin BEM to 2-D problems. *International Journal for Numerical Methods in Engineering*, **40**(11), pp. 1977–1999, 1997.

BOUNDARY ELEMENT METHOD ANALYSIS OF BOUNDARY VALUE PROBLEMS WITH PERIODIC BOUNDARY CONDITIONS

VASYL I. GNITKO[1], ARTEM O. KARAIEV[2*], NEELAM CHOUDHARY[3] & ELENA A. STRELNIKOVA[1,2†]
[1]A. Podgorny Institute of Mechanical Engineering Problems of the Ukrainian Academy of Sciences, Ukraine
[2]V.N. Karazin Kharkiv National University, Ukraine
[3]Bennett University, India

ABSTRACT

This paper explores non-axisymmetric boundary value problems for the Laplace equation. Neumann's, Dirichlet's and mixed boundary conditions are involved, supposing their periodic behaviour. Boundary value problems arise as auxiliary issues in many practical applications. Among them there are problems related to numerical simulation of vibrations of fluid-filled elastic shells of revolution, coupled vibrations of elastic circular plates resting on a sloshing liquid, crack propagation in elastic mediums, and more. The common feature in these problems is the necessity to obtain the numerical solution of the Laplace equation under different boundary conditions. As these problems are auxiliary, it is necessary to obtain their numerical solutions with high accuracy. The most effective method to solve these problems is the boundary elements method (BEM). Here a new variant of BEM is proposed for the axisymmetric calculation domain with given periodic functions for boundary conditions. The shape of the calculation domain allows us to reduce surface integral equations to one-dimensional ones. In doing so, we must evaluate elliptic-like inner integrals with high accuracy, to elaborate the method of calculation of the outer integrals with logarithmic, Cauchy or Hadamard finite part singularities. An efficient method for evaluating elliptic-like integrals was developed using a special series for integrands, and the quadrature equations were obtained for high-precision calculation of outer integrals. The method developed can be used to determine free vibration modes and frequencies for elastic fluid-filled shells of revolution.

Keywords: boundary element method, periodic boundary conditions, singular integral equations, free vibrations, fluid-filled elastic shells.

1 INTRODUCTION

Mixed boundary value problems for elliptic equations with periodic boundary conditions occur in a wide range of engineering applications, such as composite and fluid mechanics [1], [2], vibrations of structural elements [3], crack propagation in elastic mediums [4], fluid–structure interactions [5], [6], cyclically symmetrical structure design [7], and more.

In modelling mechanical processes, systems of differential equations with periodic boundary value conditions are usually involved. Analytical solutions of these equations have been obtained for some simple cases, but advanced numerical methods are currently needed in mechanical and engineering applications [8], [9]. For successive applications of numerical methods to solve periodic boundary value problems (PBVP), theorems are needed regarding periodic behaviour of the solutions. Some of these theorems were proven [10], [11]. The existing theorems allow us to use Fourier series coupled with the finite element method (FEM), finite difference method (FDM) or boundary element method (BEM), widely used in PBVP for differential equations. In using FEM and FDM, the main obstacle is mesh-generating for an arbitrary domain that requires large amounts of

* *ORCID: https://orcid.org/0000-0003-3176-8496*
† *ORCID: http://orcid.org/0000-0003-0707-7214*

WIT Transactions on Engineering Sciences, Vol 131, © 2021 WIT Press
www.witpress.com, ISSN 1743-3533 (on-line)
doi:10.2495/BE440031

processing power and computation time. In recent decades, the BEM originated and developed by Carlos Brebbia [12] has become increasingly popular, due to its unique feature of mesh reduction.

In this paper, the non-axisymmetric periodic boundary value problems of free liquid surface vibrations in shells of revolution and elastic shell vibrations are formulated, and then solved using BEM coupled with Fourier series.

2 NON-AXISYMMETRIC PERIODIC BOUNDARY VALUE PROBLEMS OF SHELL VIBRATIONS

2.1 Problem statement

Consider a coupled problem of dynamic behaviour of an elastic shell of revolution partially filled with a liquid. Such shells can be used as models for numerical simulation of liquid disturbance/sloshing and vibrations of the shell walls in fuel tanks, oil storage reservoirs, water storage containers, etc. The wet part of the shell surface is denoted as S_1 and the liquid free surface as S_0 (Fig. 1).

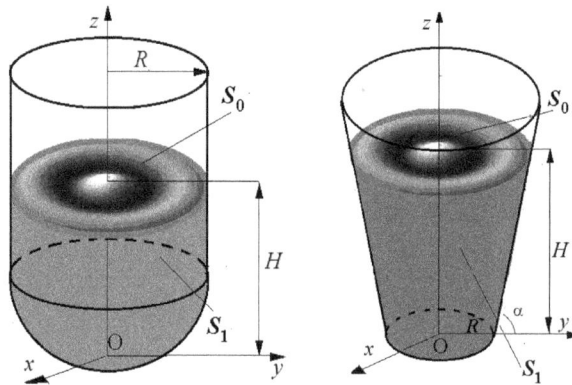

Figure 1: Shells of revolution partially filled with a liquid.

For a liquid that is flowing, we consider two conservation laws: conservation of mass and conservation of momentum. Letting $\mathbf{V}$ be the velocity field of the fluid, ρ the liquid density, the continuity equation (law of mass conservation) in the absence of mass sources is (eqn 1):

$$\frac{\partial \rho}{\partial t} + \mathrm{div}(\rho \mathbf{V}) = 0 \, .$$

The momentum conservation law states that the rate of momentum change is equal to the applied forces. Forces acting on the liquid control volume are the integral of the stress tensor, σ_{ij}, over the surface; plus the integral of body force vectors per unit mass, over the volume. Considering the motion of a continuous, viscous fluid, the stress in the fluid is composed of two parts; a locally isotropic part proportional to the scalar pressure field and a non-isotropic part, due to viscous friction. The stress tensor follows eqn (2);

$$\sigma_{ij} = -p\delta_{ij} + \tau_{ij} \,,$$

where p is the pressure, δ_{ij} is the Kronecker unit tensor, and τ_{ij} is the viscous stress tensor.

Suppose that the liquid inside the shell is an ideal $\left(\tau_{ij} = 0\right)$ one, and the surface tension is neglected: We then have the following eqn (3) for motion under the force of gravity $\rho\mathbf{g}$:

$$\rho\left(\frac{\partial \mathbf{V}}{\partial t} + (\mathbf{V} \cdot \nabla)\mathbf{V}\right) = -\nabla p - \rho\nabla(gz).$$

For incompressible inhomogeneous liquids, conservation laws in linear formulation became:

$$\operatorname{div}(\mathbf{V}) = 0 \,, \tag{1}$$

$$\rho\frac{\partial \mathbf{V}}{\partial t} = -\nabla p - \rho\nabla(gz). \tag{2}$$

If the fluid flow is not rotational ($\operatorname{rot} \mathbf{V} = 0$), then the potential flow theory could be applied. Thus, a scalar velocity potential of $\Phi = \Phi(x, y, z, t)$ can be introduced, with $\mathbf{V} = \nabla\Phi$. Then eqn (1) is transformed to the Laplace equation:

$$\Delta\Phi = 0 \,. \tag{3}$$

From momentum conservation law (2) we obtain:

$$p - p_0 = -\rho\left(\frac{\partial\Phi}{\partial t} + gz\right), \tag{4}$$

where p_0 is atmospheric pressure.

To determine the potential Φ, a mixed boundary value problem for the Laplace equation was formulated. The non-penetration condition of the wet tank surface (S_1) was applied [13]. On the free surface, the following kinematic and dynamic boundary conditions are satisfied [6], [9]:

$$\left.\frac{\partial\Phi}{\partial n}\right|_{S_0} = \frac{\partial\zeta}{\partial t}, \quad p - p_0|_{S_0} = 0 \,. \tag{5}$$

Here, an unknown function $\zeta = \zeta(x, y, t)$ is the free surface elevation measured vertically, above the still water level.

This function describes the shape and position of the free surface. Thus, the following boundary value problem was formulated:

$$\Delta\Phi = 0 \,, \quad \left.\frac{\partial\Phi}{\partial n}\right|_{S_1} = 0 \,, \quad \left.\frac{\partial\Phi}{\partial n}\right|_{S_0} = \frac{\partial\zeta}{\partial t}, \quad p - p_0|_{S_0} = 0 \,, \tag{6}$$

where $p - p_0$ is received from eqn (4) at $z = \zeta(x, y, t)$, and n is an external unit normal to the corresponding surfaces.

The solvability condition for the Neumann problem (6) is given [13]:

$$\iint\limits_{S_0} \frac{\partial \Phi}{\partial n} dS_0 = 0. \tag{7}$$

The boundary value problem (6), (7), was formulated for studying the liquid vibrations within the shell. Problems involving axisymmetric bodies with arbitrary boundary conditions are best treated using cylindrical coordinates. With a cylindrical coordinates system (ρ, z, α), one can represent the velocity potential Φ that satisfies a given boundary value problem (6), (7), as follows [6]:

$$\Phi = \sum_{l=0}^{N} \sum_{k=1}^{M} \dot{d}_{lk}(t) \varphi_{lk}(\rho, z, \alpha). \tag{8}$$

From eqns (6) and (8), the next presentation is obtained for determining the free surface elevation ζ:

$$\zeta = \frac{1}{g} \sum_{l=0}^{N} \sum_{k=1}^{M} d_{lk}(t) \zeta_{lk}(\rho, 0, \alpha). \tag{9}$$

Here, $d_{lk}(t)$ are unknown time-dependent functions, while the φ_{lk} functions satisfy the following equations:

$$\Delta \varphi_{lk} = 0, \quad \left. \frac{\partial \varphi_{lk}}{\partial n} \right|_{S_1} = 0, \quad \left. \frac{\partial \varphi_{lk}}{\partial n} \right|_{S_0} = \zeta_{lk}. \tag{10}$$

Considering fluid-filled elastic shells without including the force of gravity [5], we arrive at a similar problem. Namely, the solution for the potential Φ in the following equation is in use:

$$\Phi = \sum_{l=0}^{N} \sum_{k=1}^{M} \dot{c}_{lk}(t) \psi_{lk}(\rho, z, \alpha).$$

Here, $c_{lk}(t)$ are unknown time-dependent functions, while functions for ψ_{lk} are obtained from the following boundary value problems [5]:

$$\Delta \psi_{lk} = 0, \quad \left. \frac{\partial \psi_{lk}}{\partial n} \right|_{S_1} = u_{lk}, \quad \psi_{lk}|_{S_0} = 0, \tag{11}$$

where the u_{lk} functions are their own modes of the empty elastic shell vibrations.

If the fluid-filled shell being considered is a shell of revolution, then dependency of the angular coordinate for unknown functions can be most conveniently described by Fourier series. Therefore, for the ζ_{lk} and u_{lk} functions, the following equations are used:

$$\zeta_{lk}(\rho, 0, \alpha) = \tilde{\zeta}_{lk}(\rho) \cos l\alpha, \quad u_{lk}(\rho, z, \alpha) = \tilde{u}_{lk}(\rho, z) \cos l\alpha.$$

Then, discretisation of the body requires only meshing the shell generator with one-dimensional boundary elements, which substantially decreases the number of unknowns. The drawback of this method is in the increasing complexity of numerical implementation, because integrals over the angle for shells of revolution must be calculated for every Fourier term. Herein, we have elaborated this effective method to calculate the angular integrals.

2.2 Periodic boundary value problems

Without loss of generality, the periodic boundary value problem statement in cylindrical coordinates can be expressed in the following form:

$$\Delta u = 0, \quad \frac{\partial u}{\partial n}\bigg|_{S_1} = f(\rho, z)\cos(l\alpha), \quad u\big|_{S_0} = 0. \tag{12}$$

Here, u is an unknown function, Δ is the Laplace operator, S_0 is the free surface (of liquid), S_1 is for the inside shell surfaces, $f(\rho, z)$ is a known boundary function, and l is an integer number.

It should be noted that the boundary conditions in problems (10)–(12) are non-axisymmetric. So we need to use this property in integral representations of the problems under consideration. To solve the Laplace equation, we used the well-known third Green's identity, which represents solutions for the Laplace equation in the following integral form:

$$2\pi u(\xi) + \oint_S q^*(\xi, \mathbf{r})u(\mathbf{r})dS(\mathbf{r}) = \oint_S u^*(\xi, \mathbf{r})\frac{\partial u}{\partial n}(\mathbf{r})dS(\mathbf{r}).$$

Here, $u^*(\xi, \mathbf{r})$ and $q^*(\xi, \mathbf{r})$ are expressed by Green's function of the Laplace equation and its normal derivative, and $S = S_0 \cup S_1$.

Hereinafter, we will suppose that the body's boundary is axisymmetric with non-axisymmetric boundary conditions; therefore, we can make a transformation from surface integral to contour in the following manner, where Γ is a generator of the surface S:

$$2\pi u(\xi) + \oint_\Gamma \rho d\Gamma(\mathbf{r})\int_0^{2\pi} q^*(\xi, \mathbf{r})u(\mathbf{r})d\alpha = \oint_\Gamma \rho d\Gamma(\mathbf{r})\int_0^{2\pi} u^*(\xi, \mathbf{r})\frac{\partial u}{\partial n}(\mathbf{r})d\alpha. \tag{13}$$

In this study, due to the boundary conditions in eqn (12), we can present the unknown function in the following form:

$$u(\mathbf{r}) = u(\rho, z)\cos(l\alpha).$$

Here $u(\rho, z)$ is an unknown axisymmetric function, yet non-axisymmetric behaviour within the problem is taken into account by multiplying this function by cosines, so that eqn (13) takes the form of:

$$2\pi u(\xi) + \int_{\Gamma_1} u(\rho, z)\rho d\Gamma(\mathbf{r})\int_0^{2\pi} q^*(\xi, \mathbf{r})\cos(l\alpha)d\alpha = \int_{\Gamma_0} \frac{\partial u}{\partial n}(\rho, z)\rho d\Gamma(\mathbf{r})\int_0^{2\pi} u^*(\xi, \mathbf{r})\cos(l\alpha)d\alpha +$$

$$+ \int_{\Gamma_1} f(\rho, z)\rho d\Gamma(\mathbf{r})\int_0^{2\pi} u^*(\xi, \mathbf{r})\cos(l\alpha)d\alpha.$$

Here, $\Gamma = \Gamma_0 \cup \Gamma_1$, Γ_0 and Γ_1 are generators of the surfaces S_0 and S_1, respectively. So the topical issue addressed here is to calculate axial integrals with Green's functions, multiplied by trigonometric functions.

2.3 Axial integrals with Green's functions and trigonometric functions

It is necessary to find a way of estimating the next integrals:

$$I_u(\rho,\rho_0,z,z_0) = \int_0^{2\pi} u^*(\xi,\mathbf{r})\cos(l\alpha)d\alpha, \quad I_q(\rho,\rho_0,z,z_0) = \int_0^{2\pi} q^*(\xi,\mathbf{r})\cos(l\alpha)d\alpha. \tag{14}$$

Here, the following Green's functions are involved:

$$u^*(\xi,\mathbf{r}) = \frac{1}{|\xi - \mathbf{r}|}, \quad q^*(\xi,\mathbf{r}) = \frac{(\xi - \mathbf{r}, \mathbf{n}(\mathbf{r}))}{|\xi - \mathbf{r}|^3}. \tag{15}$$

By using the expressions described in eqn (15), one can obtain $I_u(\rho,\rho_0,z,z_0)$ as seen in the next expression:

$$\int_0^{2\pi} u^*(\xi,\mathbf{r})\cos(l\alpha)d\alpha = \int_0^{2\pi} \frac{\cos(l\alpha)d(\alpha - \alpha_0)}{\sqrt{\rho^2 + \rho_0^2 + (z - z_0)^2 - 2\rho\rho_0\cos(\alpha - \alpha_0)}}.$$

Then, with substitution of:

$$\theta = \frac{\alpha - \alpha_0 - \pi}{2} \Rightarrow \alpha - \alpha_0 = 2\theta + \pi,$$

the integral $I_u(\rho,\rho_0,z,z_0)$ transforms into the following:

$$I_u(\rho,\rho_0,z,z_0) = \frac{4(-1)^l \cos(l\alpha_0)}{\sqrt{(\rho+\rho_0)^2 + (z-z_0)^2}} \int_0^{\frac{\pi}{2}} \frac{\cos(2l\theta)d\theta}{\sqrt{1 - m^2 \sin^2\theta}}, \quad m^2 = \frac{4\rho\rho_0}{(\rho+\rho_0)^2 + (z-z_0)^2}.$$

To calculate this integral, we need to expand the cosine functions into a series. This procedure can be accomplished by using De Moivre's equation:

$$\cos(2l\theta) = \mathrm{Re}\left[e^{2il\theta}\right] = \mathrm{Re}(\cos(2l\theta) + i\sin(2l\theta)) = \mathrm{Re}(\cos\theta + i\sin\theta)^{2l} =$$

$$= \mathrm{Re}\sum_{k=0}^{2l} C_{2l}^k \cos^{2l-k}2\theta \sin^k 2\theta (i)^k = \sum_{k=0}^{2l} C_{2l}^k \cos^{2l-k}2\theta \sin^k 2\theta \, \mathrm{Re}(i)^k.$$

It should be noted that $\mathrm{Re}(i)^k = 0, \forall k = 2n+1$.

Thus, by changing the summation indexes, it is possible to obtain:

$$\sum_{k=0}^{2l} C_{2l}^k \cos^{2l-k} 2\theta \sin^k 2\theta \operatorname{Re}(i)^k = \begin{vmatrix} 2p = k \\ k=0 \Rightarrow p=0, k=2l \Rightarrow p=l \end{vmatrix} =$$

$$= \sum_{p=0}^{l} (-1)^p C_{2l}^{2p} \cos^{2(l-p)} \theta \sin^{2p} \theta = \sum_{p=0}^{l} (-1)^p C_{2l}^{2p} (1-\sin^2\theta)^{l-p} \sin^{2p} \theta =$$

$$= \sum_{p=0}^{l} (-1)^p C_{2l}^{2p} \sin^{2p} \theta \sum_{k=0}^{l-p} C_{l-p}^k (-1)^k \sin^{2k} \theta = \begin{vmatrix} s = p+k \\ k=0 \Rightarrow s=p \\ k=l-p \Rightarrow s=l \end{vmatrix} =$$

$$= \sum_{p=0}^{l} \sum_{s=p}^{l} (-1)^s C_{2l}^{2p} C_{l-p}^{l-s} \sin^{2s} \theta$$

If we then add the term $\sum_{s=0}^{p-1} (-1)^s C_{2l}^{2p} C_{l-p}^{l-s} \sin^{2s} \theta$ into the inner sum, the above result does

not change due to the properties of the Euler's gamma function. We can therefore get an expression with independent sum indexes:

$$\sum_{p=0}^{l} \sum_{s=p}^{l} (-1)^s C_{2l}^{2p} C_{l-p}^{l-s} \sin^{2s} \theta = \sum_{p=0}^{l} \sum_{s=0}^{l} (-1)^s C_{2l}^{2p} C_{l-p}^{l-s} \sin^{2s} \theta =$$

$$= \sum_{s=0}^{l} (-1)^s \sin^{2s} \theta \sum_{p=0}^{l} C_{2l}^{2p} C_{l-p}^{l-s}.$$

The inner sum can be calculated using the Yegorychev summation method [14]:

$$\sum_{p=0}^{l} C_{2l}^{2p} C_{l-p}^{l-s} = \sum_{p=0}^{s} C_{2l}^{2p} C_{l-p}^{l-s} = \frac{4^s l}{l+s} C_{l+s}^{2s}.$$

Thus, the double cosine function has the following presentation:

$$\cos(2l\theta) = \sum_{s=0}^{l} \frac{4^s (-1)^s l}{l+s} C_{l+s}^{2s} \sin^{2s} \theta.$$

Using expansion in eqn (16), the integral $I_u(\rho, \rho_0, z, z_0)$ was transformed into:

$$I_u(\rho, \rho_0, z, z_0) = \frac{4(-1)^l \cos(l\alpha_0)}{\sqrt{(\rho+\rho_0)^2 + (z-z_0)^2}} \sum_{k=0}^{l} \frac{(-1)^k 4^k l}{k+l} C_{l+k}^{2k} \int_0^{\frac{\pi}{2}} \frac{\sin^{2k} \theta \, d\theta}{\sqrt{1-m^2 \sin^2 \theta}}. \tag{16}$$

The squares of the sine functions can be expanded into the following series:

$$\sin^{2k} \theta = \frac{1}{m^{2k}} \sum_{s=0}^{k} (-1)^s C_k^s (1-m^2 \sin^2 \theta)^s.$$

So we get eqn (17):

$$\int_0^{\frac{\pi}{2}} \frac{\sin^{2k}\theta \, d\theta}{\sqrt{1-m^2\sin^2\theta}} = \frac{1}{m^{2k}}\sum_{s=0}^{k}(-1)^s C_k^s \int_0^{\frac{\pi}{2}}\left(1-m^2\sin^2\theta\right)^{\frac{2s-1}{2}}d\theta \, . \tag{17}$$

Finally, in order to evaluate the integral $I_u(\rho,\rho_0,z,z_0)$ we used equations given in eqns (16) and (17).

In reference [15], the special elliptic integrals are introduced:

$$E_s(m) = \int_0^{\frac{\pi}{2}}\left(1-m^2\sin^2\theta\right)^{\frac{2s-1}{2}}d\theta \, . \tag{18}$$

The simplest approach for calculation of the integrals in eqn (18) is by using numeric integration, but increasing the order s of integrals, so numerical error also grows. In reference [15], the recurrent equations are in use. The special elliptic integrals of order s are expressed as:

$$E_s(m) = Q_s(m)E(m) + P_s(m)K(m) \, . \tag{19}$$

Here, $E(m)$ and $K(m)$ are complete elliptic integrals of the first and second kind, and $Q_s(m)$, $P_s(m)$ are polynomials. In previous work [15], [16], the method for high-accuracy calculations of complete elliptic integrals of the first and second kind is developed using the well-known arithmetic-geometric mean (AGM), Gauss' technique [17].

The analogical procedure is elaborated here for evaluating the integral $I_q(\rho,\rho_0,z,z_0)$. Here the relation for eqn (19) is also obtained, but with other polynomials $Q_s(m)$ and $P_s(m)$.

After calculating the inner integrals $I_u(\rho,\rho_0,z,z_0)$ and $I_q(\rho,\rho_0,z,z_0)$ in eqn (13), the boundary element method (BEM) with constant approximation of densities was applied to obtain unknown functions using a special approach, to calculate the outer integrals with logarithmical singularities [2], [5].

3 NUMERICAL SIMULATIONS

3.1 Benchmark test

For testifying the proposed numerical algorithm with qualified evaluation of inner integrals the liquid frequencies in the rigid cylindrical shell are compared with R. Ibrahim's analytical solutions (eqn 20), as seen in previously published work [18]:

$$\frac{\chi_k^2}{g} = \frac{\mu_k}{R}\tanh\left(\mu_k\frac{H}{R}\right), k=1,2,..\, ; \varphi_k = J_l\left(\frac{\mu_k}{R}r\right)\cosh\left(\frac{\mu_k}{R}z\right)\cosh^{-1}\left(\frac{\mu_k}{R}H\right). \tag{20}$$

Here, R is the shell radius, H is its height, μ_k values are roots of the equation $J_l'(x)=0$, the $J_l(x)$ is a Bessel function of the first kind; and χ_k, φ_k are the frequencies and modes of liquid sloshing within the rigid cylindrical shell. The numerical solutions are obtained by using the BEM as it was described beforehand, as well as by the method developed in reference [8].

Suppose that $R = 1$ m and $H = 1$ m. Table 1 provides the numerical values of the natural frequencies of liquid sloshing for $l = 1$, obtained by the proposed numerical method, the method developed in reference [8], and the analytical solution for eqn (20). Thus, Table 1 results testify to the convergence of our proposed BEM and its effectiveness. Here, in both numerical methods, we considered equal numbers of $N = 150$ for the boundary elements along Γ.

Table 1: Slosh frequency parameters χ_n^2 / g of a fluid-filled rigid cylindrical shell, using different solving methods.

Method	$n = 1$	$n = 2$	$n = 3$	$n = 4$	$n = 5$
[8]	1.6590	5.3301	8.5385	11.7071	14.8684
Proposed BEM	1.6573	5.3293	8.5363	11.7060	14.8635
Analytical solution (20)	1.6573	5.3293	8.5363	11.7060	14.8635

The results obtained by both numerical methods were very close to the analytical ones. It should be noted that the results obtained by our new variant of BEM, with qualified evaluation of inner integrals, are essentially more accurate.

3.2 Vibrations of elastic fluid-filled truncated conical shells

We also provide estimation of natural modes and frequencies of an elastic truncated cone coupled with liquid sloshing. In Table 2, an elastic conical tank with clamped-free edges was considered, supposing that $R = 0.5$ m, $H = 1$ m and $\theta = \pi/4$. Results obtained were for seven wave numbers, $l = \overline{0.6}$ with $k = \overline{1.4}$. The frequencies of liquid sloshing, vibrations of empty and liquid-filled tanks are obtained. The results of numerical simulations are given in Table 2.

For these numerical simulations, three basic systems were built, similar to previous studies [6]. The elastic empty shell modes are first. The displacements in the coupled problem are considered as the linear combinations of the empty elastic shell natural modes. Free vibration modes and frequencies for the liquid-filled elastic shell, without gravity effects, are defined at the second step. These are the second system of basic functions. The third system consists of sloshing modes, including gravity effects. The numerical simulation is accomplished using the developed one-dimensional BEM with qualified evaluation of the inner integrals.

Hereinafter, in numerical simulation, the shell thickness and Poisson's ratio were taken as $h/R = 0.01$ and $v = 0.3$, Young's modulus was $E = 2.11 \times 10^6$ MPa, and densities for shells and liquid are $\rho_s = 8,000$ kg/m^3 and $\rho_l = 1,000$ kg/m^3, respectively.

In this case, the mutual influence of sloshing and elastic shell vibrations is negligible. The separation of frequency spectra for the liquid-filled elastic cone tank and the liquid sloshing within the rigid tank was observed.

The conclusions following numerical simulations are interesting: First, the lowest frequency here belongs to the axisymmetric mode with dominant bottom vibrations. The axisymmetric modes for the bottom and the walls of the truncated conical tanks are demonstrated in Figs 2 and 3.

Table 2: Frequencies in elastic truncated conical shell, in Hertz (Hz).

n	m	Frequencies		
		Sloshing	Empty	Fluid-filled
0	1	5.836	101.07	41.67
	2	8.300	393.49	214.06
	3	9.997	559.52	257.91
	4	11.443	675.88	471.43
1	1	3.659	210.34	113.56
	2	7.001	327.90	126.64
	3	8.979	601.83	425.00
	4	10.577	649.99	438.22
2	1	4.819	193.05	96.57
	2	7.897	345.07	224.03
	3	9.729	605.43	346.52
	4	11.236	764.39	500.66
3	1	5.707	128.30	64.217
	2	8.661	504.88	281.69
	3	10.397	519.84	327.95
	4	11.837	723.61	474.97
4	1	6.460	100.89	58.200
	2	9.340	436.93	467.25
	3	11.005	689.26	265.86
	4	12.394	693.20	506.67
5	1	7.1288	101.85	56.908
	2	9.9581	385.15	232.28
	3	11.568	671.56	452.47
	4	12.915	897.75	686.67
6	1	7.736	123.20	78.861
	2	10.529	368.32	241.18
	3	12.094	663.03	458.88
	4	13.406	952.83	688.34

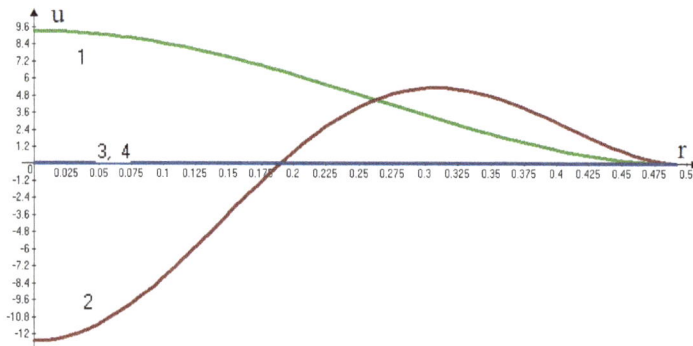

Figure 2: Axisymmetric modes of tank bottom vibrations.

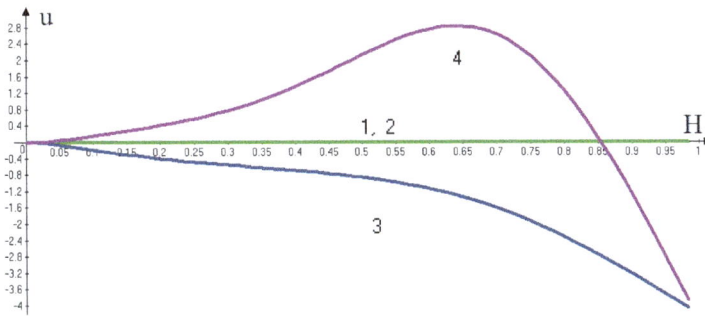

Figure 3: Axisymmetric modes of tank wall vibrations.

Hereinafter, the numbers 1, 2, 3 and 4 correspond to the vibration numbers for k with $l = 0$. Figs 2 and 3 demonstrate different behaviour for the bottom and for the shell walls' vibrations. One can observe that in this case, the bottom and wall vibrations do not affect each other. Note that the frequency $\omega = 41.67$ Hz is the lowest one for vibrations of the liquid-filled elastic conical shell with elastic bottom. It corresponds to $l = 0$ and $k = 1$. If conical shells with rigid bottoms are considered, then the lowest frequency occurs at $l = 4$ and $k = 1$ for the empty shell; and $l = 5$ and $k = 1$ for the fluid-filled shell. Fig. 4 demonstrates that modes corresponding to the bottom vibrations are $l = 5$ and $k = 1$.

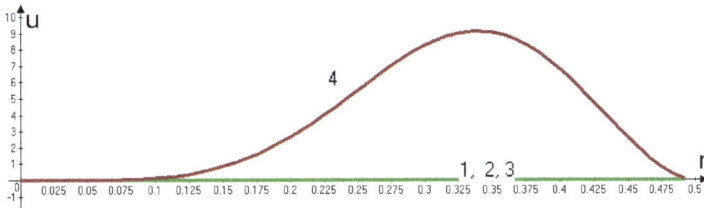

Figure 4: Modes of tank bottom vibrations given $l = 5$ and $k = 1$.

Fig. 5 demonstrates the wall vibration modes for $l = 5$ and $k = 1$. The lowest frequency here responds to the wall vibrations. Thus, if the truncated elastic conical tank with a rigid bottom is considered, then the lowest frequency does not correspond to the axisymmetric mode.

Analysis of Figs 2–5 should lead to an understanding that at low wave numbers, the dominant modes of truncated elastic cone vibrations correspond to its bottom, yet with increasing the number of nodal diameters, the tank's wall vibrations become dominant.

Fig. 6 shows the modes of lowest frequencies for liquid sloshing in the rigid tank (left); plus for the elastic conical tank with the rigid bottom (right).

If the bottom deformation is neglected, then the lowest frequency of elastic fluid-filled shell would be missed.

The frequencies (ω) near 100 Hz were considered as being most dangerous for empty shells. Our results, provided in Table 2, testify to it. For example, a $\omega = 101.07$ Hz corresponds to $l = 0$ and $k = 1$; and a $\omega = 100.89$ Hz corresponds to an $l = 4$ and $k = 1$; plus a $\omega = 101.86$ Hz corresponds to $\underline{l} = 5$ and $k = 1$.

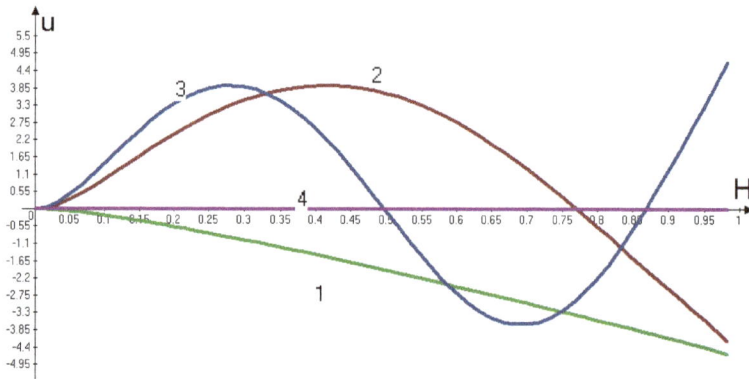

Figure 5: Modes of tank wall vibrations for $l = 5$ and $k = 1$.

Figure 6: Modes of lowest frequencies.

It is also important to note that the lowest frequencies of the empty and liquid-filled tanks corresponded to different circumferential wave numbers.

The frequencies of liquid-filled tank vibrations are drastically different from the frequencies of empty tanks; however, with increasing the wave number, this difference became gradually smaller.

4 CONCLUSIONS

We propose a new variant of BEM that solves the Laplace equation with axisymmetric calculation domain, giving periodic functions under boundary conditions. The shape of the calculation domain allowed us to reduce surface integral equations down to one-dimensional ones. We introduced special elliptic integrals and developed an advanced numerical method for their evaluation. We analysed the natural vibrations for truncated elastic shell interactions with internal liquid sloshing/rotating within shell forms. Coupled one-dimensional finite and boundary element methods were used. The vibration analysis included several steps. The numerical simulation was accomplished using the developed one-dimensional BEM with qualified evaluation of inner integrals. The developed method essentially reduces computer time for numerical analysis and produces new qualitative possibilities in advanced computational modelling of the dynamic characteristics of liquid-filled elastic shell structures. We demonstrated the differences in dynamic characteristics between elastic truncated shells with rigid and elastic bottoms. Our results can be considered as a basis for further research in the dynamic behaviour of structures subjected to intensive loading and fluid interactions. It should be noted that our interpretation and understanding of the dynamic processes found in elastic shell structures subjected to the actions of flowing fluids is far from completion and requires additional research.

ACKNOWLEDGEMENTS

We gratefully acknowledge financial support for our project from the Ministry of Education and Science of Ukraine. The authors would also like to thank our foreign collaborator, Professor Alexander Cheng, at the University of Mississippi, USA, for his constant support and interest to our research.

REFERENCES

[1] Sadaba, S., Herraez, M., Naya, F., Gonzalez C., Llorca, J. & Lopes, C.S., Special-purpose elements to impose periodic boundary conditions for multiscale computational homogenization of composite materials with the explicit finite element method. *Composite Structures*, **193**, pp. 1–24, 2018. DOI: 10.1016/j.compstruct.2018.10.037.

[2] Gnitko, V., Degtyariov, K., Karaiev, A. & Strelnikova, E., Multi-domain boundary element method for axisymmetric problems in potential theory and linear isotropic elasticity. *WIT Transactions in Engineering Sciences*, vol. 122, WIT Press: Southampton and Boston, pp. 13–25, 2019. DOI: 10.2495/BE410021.

[3] Rao, L.B. & Rao, C.K., Free vibration of circular plates with elastic edge support and resting on an elastic foundation. *International Journal of Acoustics and Vibration*, **17**(4), pp. 204–207, 2012.

[4] Liu, C., Yang, M., Han, H. & Yue, W., Numerical simulation of fracture characteristics of jointed rock masses under blasting load. *Engineering Computation*, **36**, pp. 1835–1851, 2019.

[5] Gnitko, V., Marchenko, U., Naumenko, V. & Strelnikova, E., Forced vibrations of tanks partially filled with the liquid under seismic load. *WIT Transactions on Modelling and Simulation*, vol. 52, WIT Press: Southampton and Boston, pp. 285–296, 2011. DOI: 10.2495/BE110251.

[6] Strelnikova, E., Choudhary, N., Kriutchenko, D., Gnitko, V. & Tonkonozhenko, A., Liquid vibrations in circular cylindrical tanks with and without baffles under horizontal and vertical excitations. *Engineering Analysis with Boundary Elements*, **120**, pp. 13–27, 2020. DOI: 10.1016/j.enganabound.2020.07.024.

[7] Misura, S., Smetankina, N. & Misiura, I., Optimal design of the cyclically symmetrical structure under static load. *Integrated Computer Technologies in Mechanical Engineering 2020*, Springer: Cham, pp. 256–266, 2020. DOI: 10.1007/978-3-030-66717-7_21.

[8] Gnitko, V., Degtyariov, K., Karaiev, A. & Strelnikova, E., Singular boundary method in a free vibration analysis of compound liquid-filled shells. *WIT Transactions on Engineering Sciences*, vol. 126, WIT Press: Southampton and Boston, pp. 189–200, 2019. DOI: 10.2495/BE420171.

[9] Smadi, M., Arqub, O.A. & Ajou, A., A numerical iterative method for solving systems of first-order periodic boundary value problems. *Journal of Applied Mathematics*, **2014**, 10 p., 2014. DOI: 10.1155/2014/135465.

[10] Lakshmikantan, V., Periodic boundary conditions of first and second order differential equations. *Journal of Applied Mathematics and Simulations*, **2**(3), pp. 131–139, 1989.

[11] Tomeček, J., Dirichlet boundary value problem for differential equation with ϕ-Laplacian and state-dependent impulses. *Mathematica Slovaca,* **67**(2), pp. 483–494, 2017. DOI: 10.1515/ms-2016-0283.

[12] Brebbia, C.A., Telles, J.C.F. & Wrobel, L.C., *Boundary Element Techniques*, Springer-Verlag: Berlin and New York, 1984.

[13] Gavrilyuk, I., Lukovsky, I., Trotsenko, Y. & Timokha, A., Sloshing in a vertical circular cylindrical tank with an annular baffle. Part 1: Linear fundamental solutions. *Journal of Engineering Mathematics*, **54**, 2006, pp. 71–88.

[14] Egorychev, G.P., Method of coefficients: An algebraic characterization and recent applications. *Advances in Combinatorial Mathematics: Proceedings of the Waterloo Workshop in Computer Algebra 2008*, Springer, Devoted to the 70th birthday of G. Egorychev, pp. 1–30, 2009.

[15] Karaiev, A. & Strelnikova, E., Axisymmetric polyharmonic spline approximation in the dual reciprocity method. *Zeitschrift für Angewandte Mathematik und Mechanik*, **101**, p. e201800339, 2021. DOI: 10.1002/zamm.201800339.

[16] Zheng, L., Gaofeng, W., Zhiming, W. & Jinwei, Q., The meshfree analysis of geometrically nonlinear problem based on radial basis reproducing kernel particle method. *International Journal of Applied Mechanics*, **12**(4), p. 2050044, 2020. DOI: 10.1142/S1758825120500441.

[17] Cox, D.A., The arithmetic-geometric mean of gauss. *L'Enseignement Mathématique*, **30**, pp. 275–330, 1984.

[18] Ibrahim, R.A., *Liquid Sloshing Dynamics. Theory and Applications*. Cambridge University Press, 972 pp., 2005.

CALCULATION OF CAUCHY-TYPE INTEGRALS NEAR CONTOURS IN DIRECT AND INVERSE ELASTIC PROBLEMS

ALEXANDER N. GALYBIN

Schmidt Institute of Physics of the Earth (IPE RAS), Russia

ABSTRACT

This study presents an approach for the calculation of Cauchy-type integrals at points located near contours. It is evident that the kernel of a Cauchy integral becomes close to singular as soon as one intends to calculate the value of the integral close to the contour. As a result, more nodes in a quadrature formula are needed, in order to reach acceptable accuracy in the calculations. This problem is faced in standard formulations when analysing stress–strain states after obtaining numerical solutions of certain singular integral equations; as well as in non-classical formulations, where the data close to the contour are used as input. On the other hand, one can employ, for the contour points, the Plemelj–Sokhotski formulas, assuming calculation of the singular integral is followed by addition of a known non-integral term. In this study, we use expansions into power series to calculate stress characteristics at points near the contours, suggest an algorithm, and numerically analyse two cases that are relevant to direct and inverse formulations in plane elasticity.

Keywords: plane elasticity, Cauchy integrals, singular integral equations, inverse problem, mathematical optimisation.

1 INTRODUCTION

This study aims to develop numerical techniques for the calculation of complex potentials near contours. It is well known that the general solution of a plane elastic problem is expressed via two independent complex potentials [1]. These can be presented by Cauchy-type integrals with unknown densities, as follows:

$$\Phi(z) = \frac{1}{2\pi i} \int_\Gamma \frac{g(t)}{t-z}\, dt, \quad \Psi(z) = \frac{1}{2\pi i} \int_\Gamma \frac{h(t)}{t-z}\, dt \,. \tag{1}$$

Here, Γ is a closed or open contour (or set of contours), and the densities $g(t)$ and $h(t)$ are not independent of each other. They are found by solving certain singular integral equations corresponding to a particular boundary value problem under consideration.

The method of mechanical quadratures [2] is often used to solve a system of complex singular integral equations (CSIE), because of its high efficiency and simplicity in programming. The sought density is found at the collocation points that depend on the quadrature formulas used, usually presenting the roots of the Chebyshev polynomials. Interpolation at other contour points can be made by different methods, such as linearly piecewise, cubic splines or by interpolation polynomials. It is worth noting that the accuracy of the interpolated solution is always not as good as the accuracy of the solution at the collocation points; that is, $2n$ where n is the order of the Chebyshev polynomials used in approximation of the sought density. The contour values of the complex potentials are found by the Plemelj–Sokhotski formulas:

$$2\Phi^{\pm}(\zeta) = \pm g(\zeta) + \mathbf{S}(g), \quad \zeta \in \Gamma, \tag{2}$$

WIT Transactions on Engineering Sciences, Vol 131, © 2021 WIT Press
www.witpress.com, ISSN 1743-3533 (on-line)
doi:10.2495/BE440041

where $\mathbf{S}(g)$ is the singular integral below:

$$\mathbf{S}(g) = \frac{1}{\pi i} \int_{\Gamma} \frac{g(t)}{t - \zeta} dt, \quad \zeta \in \Gamma. \tag{3}$$

Numerical methods for the calculation of singular integrals (3) are well developed. For instance, if Γ is an open interval $(-1,1)$, then for unbounded solutions at the ends, the following quadrature formula can be used [2]:

$$\int_{-1}^{1} \frac{u(t)}{\sqrt{1-t^2}\,(t-x)} dt = \frac{\pi}{n} \sum_{k=1}^{n} \frac{u(\xi_k)}{\xi_k - x} + \pi u(x) \frac{U_{n-1}(x)}{T_n(x)}, \quad |x| < 1. \tag{4}$$

Here $T_n(x) = \cos(n \arccos x)$, $U_{n-1}(x) = \sin(n \arccos x)/\sqrt{1-x^2}$ are the Chebyshev polynomials of the first and second order, respectively; while the nodes ξ_k are the roots of $T_n(x)$, such as: $\xi_k = \cos \frac{2k-1}{2n} \pi$, $k = 1,\dots n$. It is evident that the second term on the right-hand side of eqn (4) disappears at the roots of $U_{n-1}(x)$, i.e., at the points $\eta_m = \cos \frac{m\pi}{n}$, $m = 1,\dots n-1$. Therefore, eqn (4) becomes:

$$\int_{-1}^{1} \frac{u(t)}{\sqrt{1-t^2}\,(t-\eta_m)} dt = \frac{\pi}{n} \sum_{k=1}^{n} \frac{u(\xi_k)}{\xi_k - \eta_m}. \tag{5}$$

This coincides with the standard quadrature formula for regular integrals. This equation is exact, if $u(t)$ is a polynomial of up to $2n$ degree.

Consider the Cauchy-type integral in eqn (1) and apply the same quadrature formula to it, which results in:

$$\int_{-1}^{1} \frac{u(t)}{\sqrt{1-t^2}\,(t-z)} dt = \int_{-1}^{1} \frac{(t-x)-iy}{(t-x)^2 + y^2} \frac{u(t)}{\sqrt{1-t^2}} dt = \frac{\pi}{n} \sum_{k=1}^{n} \frac{(\xi_k - x)-iy}{(\xi_k - x)^2 + y^2} u(\xi_k), \quad |x| < 1. \tag{6}$$

This equation is exact if the integrand (without the weight) is a polynomial of up to $2n-1$ degree. We can see that for small y the denominator in eqn (6) becomes close to singular, and so cannot be accurately approximated by the polynomials. In practice, it means that for acceptable accuracy, one would need to use approximations using polynomials of large degrees; thus, a large number of nodes and collocation points would be required. It can be estimated that for the values of y smaller than the minimum difference of $\xi_k - \eta_m$ there would be strongly degraded accuracy of eqn (6).

We aimed to develop simple algorithms that do not require the use of a large number of nodes for the calculation of integrals in eqn (1), specifically for the points located close to the contours. Our main idea was to use the properties of holomorphic functions and Taylor's expansion near the contours.

2 INTEGRAL EQUATIONS FOR USE IN DIRECT AND INVERSE PROBLEMS

General solutions for plane elastic problems in terms of complex potentials (1) are given by the Kolosov–Muskhelishvili equations for the stress and displacement components in a Cartesian coordinate system Oxy, as follows [1]:

$$P \equiv \frac{\sigma_{xx} + \sigma_{yy}}{2} = \Phi(z) + \overline{\Phi(z)},$$

$$D \equiv \frac{\sigma_{yy} - \sigma_{xx}}{2} + i\sigma_{xy} = \overline{z}\Phi'(z) + \Psi(z), \qquad (7)$$

$$2GW \equiv 2G(u_x + iu_y) = \kappa\phi(z) - z\overline{\Phi(z)} - \overline{\psi(z)}.$$

Here, $\sigma_{xx}, \sigma_{yy}, \sigma_{xy}$ are stress components; P is mean stress, D is the complex stress deviator, $W = u_x + iu_y$ is complex displacement vector with the components u_x, u_y along the x and y axes, respectively. Also, G is the shear modulus, $\kappa = (3–4\nu)$ for the plain stress and $\kappa = (3 - \nu)/(1 + \nu)$ for the plane stress conditions, where ν is Poisson's ratio, and $\Phi(z) = \phi'(z)$, $\Psi(z) = \psi'(z)$ are complex potentials (holomorphic functions of a complex variable $z = x + iy$).

By making use of eqns (1)–(3), we find the following expressions for contour values of stress functions P and D:

$$2P^{\pm} = \pm(g + g') + \mathbf{S}(g) + \overline{\mathbf{S}(g)} \qquad (8)$$

$$2D^{\pm} = \pm(h - e^{-2i\vartheta}g) + \mathbf{S}(h - e^{-2i\vartheta}g) - \mathbf{S}((\bar{t} - \bar{\zeta})g') \qquad (9)$$

Here, ϑ is the angle between the positive direction of the real axis and the tangent to the contour. For the limiting values of the stress vector $\sigma(\zeta) = \sigma_n(\zeta) + i\sigma_t(\zeta)$ $\left(\sigma(\zeta) = P(\zeta) + e^{-2i\vartheta(\zeta)}\overline{D(\zeta)}\right)$ and the tangential derivative of the displacement vectors $W'(\zeta) = \kappa\Phi(\zeta) - \overline{\Phi(\zeta)} - e^{-2i\vartheta(\zeta)}\left(\zeta\overline{\Phi'(\zeta)} - \overline{\Psi(\zeta)}\right)$ one finds

$$2\sigma^{\pm} = \pm(g + e^{-2i\vartheta}h) + \mathbf{S}(g) + \overline{\mathbf{S}(g)} + e^{-2i\vartheta}\overline{\mathbf{S}(h - e^{-2i\vartheta}g)} - e^{-2i\vartheta}\overline{\mathbf{S}((\bar{t} - \bar{\zeta})g')} \quad (10)$$

$$2W'^{\pm} = \pm(\kappa g - e^{-2i\vartheta}h) + \kappa\mathbf{S}(g) - \overline{\mathbf{S}(g)} + e^{-2i\vartheta}\overline{\mathbf{S}(h - e^{-2i\vartheta}g)} - e^{-2i\vartheta}\overline{\mathbf{S}((\bar{t} - \bar{\zeta})g')} \quad (11)$$

Now we can derive the following CSIE for the first and the second fundamental problems of plane elasticity. Assuming that the known stress vector $N + iT$ is continuous across the contour, one finds $h = -e^{-2i\vartheta}\overline{g}$ and thus, from eqn (10) we find:

$$\mathbf{S}(g) + \overline{\mathbf{S}(g)} - e^{-2i\vartheta}\overline{\mathbf{S}(e^{-2i\vartheta}(g + \overline{g}))} - e^{-2i\vartheta}\overline{\mathbf{S}((\bar{t} - \bar{\zeta})g')} = 2N + 2iT \qquad (12)$$

In the second fundamental problem, we assumed that the known displacement vector $W'_x + iW'_y$ is continuous; thus, $\kappa g = e^{-2i\vartheta}\overline{h}$ and so we get:

$$\kappa \mathbf{S}(g) - \overline{\mathbf{S}(g)} + e^{-2i\vartheta}\overline{\mathbf{S}(e^{-2i\vartheta}(\kappa \overline{g} - g))} - e^{-2i\vartheta}\overline{\mathbf{S}((\overline{t} - \overline{\zeta})g')} = 2W'_x + 2iW'_y \quad (13)$$

It can be shown that these CSIE coincide with previously published CSIE [3]. It is evident that both are of the first kind, so they can be solved by the method of mechanical quadratures outlined above. As a result, one finds the density $g(t)$, as well as $h(t)$. Thus, both potentials in eqn (1) become known. However, in order to analyse the stress–strain state in the entire domain, one should perform integration of the complex potentials, to obtain the stress functions and stress components, which implies integration at the near-contour points.

Now let us discuss a previously-considered non-classical formulation [4]. The problem was formulated as follows: Find the complex potentials by using the data on the direction of the maximum principal stress (known angle θ_j counted from the real axis), at a number of discrete points z_j ($j = 1...N$), given inside the domain (these may also be on the boundary).

The principal directions are determined by the complex stress deviator as follows:

$$D = \tau_{max}e^{i\alpha}, \quad \tau_{max} = |D| = \frac{\sigma_1 - \sigma_2}{2} \geq 0, \quad \alpha = \arg(D) = \pi - 2\theta \quad (-\pi < \alpha \leq \pi). \ (14)$$

Here, σ_1 and σ_2 are the principal stresses, θ is the direction of the major principal stress, σ_1, τ_{max} is the maximum shear stresses. Bearing in mind that τ_{max} is real, one can write:

$$\text{Im}\left[D(z,\overline{z})e^{2i\theta(z,\overline{z})}\right] = 0. \quad (15)$$

So far as the principal directions are known at points z_j one arrives at the following optimisation problem:

$$\sum_{j=1}^{N}\left\{\text{Im}\left[\exp(-i\alpha_j)\left(\overline{z}_j\Phi'(z_j) + \Psi(z_j)\right)\right]\right\}^2 \Rightarrow \min. \quad (16)$$

Let us seek the complex potentials in eqn (1), assuming continuity of the stress vector across the closed contour Γ (similar to the first fundamental problem in eqn (12)), and introduce the real and imaginary parts of the sought function:

$$\mu(t) = \text{Re}\left(g(t)\right), \quad v(t) = \text{Im}\left(g(t)\right). \quad (17)$$

Then, the complex potentials assume the following form:

$$\Phi(z) = \frac{1}{2\pi i}\int_{\Gamma}\frac{\mu(t) + iv(t)}{t - z}\,dt, \quad \Psi(z) = \frac{-1}{2\pi i}\int_{\Gamma}\frac{2\mu(t)e^{-2i\vartheta(t)} + \overline{t}\left(\mu'(t) + iv'(t)\right)}{t - z}\,dt. \quad (18)$$

Thus, the expression for the complex stress deviator is:

$$D(z,\overline{z}) = \frac{-1}{\pi i}\int_{\Gamma}\frac{\mu(t)e^{-2i\theta(t)}}{t - z}\,dt - \frac{1}{2\pi i}\int_{\Gamma}\frac{\overline{t} - \overline{z}}{t - z}\left(\mu'(t) + iv'(t)\right)dt. \quad (19)$$

Expressions (18) and (19) should satisfy the conditions of the single valuedness of the displacements:

$$\int_{\Gamma} \left[\mu(t) + i v(t) \right] dt = 0 .$$ (20)

The boundary values of eqn (19) are to be found using eqns (2) and (3), as follows:

$$D^{\pm} = \mp e^{-2i\vartheta} \mu - S(e^{-2i\vartheta} \mu) + \frac{1}{2} \mathbf{R}_1 (\mu + i v)$$ (21)

Here, the regular integral $\mathbf{R}_1$ appeared due to the second integral on the right-hand side of eqn (19), which is non-singular:

$$\mathbf{R}_1(g) = -S\!\left((\bar{t} - \bar{\zeta}) g' \right) = -\frac{1}{\pi i} \int_{\Gamma} \frac{\bar{t} - \bar{\zeta}}{t - \zeta} g'(t)\, dt = \frac{1}{\pi i} \int_{\Gamma} \left(\frac{d\bar{t}/dt}{t - \zeta} - \frac{\bar{t} - \bar{\zeta}}{(t - \zeta)^2} \right) g(t)\, dt.$$ (22)

In the optimisation problem (16), one must use the expression for the deviator in eqn (19) for internal points and in eqn (21) for boundary points. As it is evident from eqn (19) for the points located close to the contour, the first integral on the right-hand side is of a form similar to eqn (6). Calculation of the second integral does not present any numerical difficulties, because its kernel is exp[−2iarg (t − ζ)], i.e., piecewise continuous and bounded.

Numerical approaches for solving eqn (16) can also be based on the method of mechanical quadratures, which implies numerical calculations of the integrals in eqns (19) and (21). If we assume that all data are internal, then the problem is reduced to a linear system of algebraic equations:

$$\begin{cases} \mathrm{Im} \left[\sum_{k=1}^{n} w_k \, \frac{2\mu_k e^{-2i\vartheta_k} - \left(e^{-2i\vartheta_k} - e^{-2i\arg(t_k - z_j)} \right) \left(\mu_k + i v_k \right)}{2\pi i \exp(-2i\theta_j)(t_k - z_j)} \right] = 0, \quad j = 1\ldots N, \\[4mm] \sum_{k=1}^{n} w_k \left(\mu_k + i v_k \right) = 0. \end{cases}$$ (23)

Here, t_k and w_k are, respectively, nodes and weights of a quadrature formula used for the discretisation, $\mu_k = \mu(t_k)$ and $v_k = v(t_k)$ are the sought real values of the density of the potential, while $\vartheta_k = \vartheta(t_k)$ are the slope angles at the nodes, θ_j are the known principal directions at the data points z_j.

The system shown in eqn (23) consists of N + 2 real equations for the determination of 2n real unknowns. It is overspecified if N + 2 > 2n; and thus, should be solved approximately using the least squares method. We point out that for experimentally acquired data subject to measurement errors, the matrix of the system is affected by these errors. In the case when a part of the data is on the boundary, the first group of equations in eqn (23) should be modified by adding non-integral terms in eqn (21), which does not change its dimensions.

3 NUMERICAL APPROACH

3.1 Algorithm

It is evident from the CSIE that we presented in the previous section, that a loss of accuracy in calculations of the complex potentials is observed for the Cauchy type integrals $\frac{1}{2\pi i}\int_{\Gamma}\frac{g(t)}{t-z}dt$ (regardless of whether the contour is closed or open), when point z is close to the contour. On the other hand, the singular integrals do not introduce any difficulties, even though in accordance with eqn (2), addition of one extra term is required. This allows us to employ Taylor's expansion of the complex potentials (1). As soon as eqn (1) presents two holomorphic functions which derivatives with respect to the conjugated variable, the Cauchy–Riemann conditions, vanish one can write:

$$\Phi(z)=\Phi(\zeta)+\Phi'(\zeta)(z-\zeta)+\bar{\bar{o}}(|z-\zeta|),\quad \zeta\in\Gamma. \qquad (24)$$

The derivative of the holomorphic function is found in the form [5]:

$$\Phi'(z)=\frac{1}{2\pi i}\int_{\Gamma}\frac{g'(t)}{t-z}dt. \qquad (25)$$

Note that this equation is valid for any smooth, closed contours, as well as for the open contours; provided that condition (20) is satisfied. Its boundary value can be found by the Plemelj–Sokhotski formulas as well as the boundary value of $\Phi(\zeta)$. Substitution of these boundary values into expression (24), leads to the following approximate equation:

$$2\Phi(z)\approx g(\zeta)+g'(\zeta)(z-\zeta)+\mathbf{S}(g)+(z-\zeta)\mathbf{S}(g'),\quad \zeta\in\Gamma. \qquad (26)$$

The proposed algorithm supposes that a specific data point z is supplied, and the equation for Γ is known. It assumes the following simple steps:

1. Find the closest distance from the point to the contour.
2. If this distance is greater than a selected threshold, then perform integration based on the usual quadrature formulas for regular integrals.
3. If the minimum distance is smaller than the threshold, then find the distance from z to the nearest node:

    a) If this distance is small enough (below the threshold), then calculate the value of the complex potential by eqn (26), using the nearest node as the contour variable ζ.
    b) If the distance is greater than the selected threshold, then use an interpolation formula to determine ζ as the point closest to z, followed by calculations using eqn (26). Interpolation of the solution is also required, due to the non-integral terms present in eqn (26).

Note that if the contour is given by a set of ordered data points, then one may need to use an interpolation procedure. Polygonal approximation seems to be the simplest way, as it also allows one to find the slopes to the contour directly. More advanced approximations can be obtained by applying smoothing (e.g., by the fast Fourier transformation technique).

3.2 Example: Stresses near a rectilinear crack

Let us consider the case of a rectilinear crack formed by a normal load, such that the density of the potential $\Phi(z)$ is bounded at the ends. For the sake of simplicity, let us assume that the potential is dimensionless (normalised by a characteristic value of applied load) and that the crack is of a unit half-length. Then it can be presented as follows:

$$\Phi(z) = \Phi_{sing}(z) + \Phi_{regul}(z), \quad \Phi_{sing}(z) = \frac{1}{2\pi} \int_{-1}^{1} \frac{\sqrt{1-t^2}}{t-z} dt , \tag{27}$$

Here, the term $\Phi_{regul}(z)$ on the right hand side remains regular, when z tends to the contour. Therefore, we can analyse the first term $\Phi_{sing}(z)$ only.

Evaluation of the Cauchy integral gives an analytical solution for $\Phi_{sing}(z)$ in the form:

$$\Phi_{sing\_anal}(z) = \frac{1}{2}\left(\sqrt{z^2-1} - z\right). \tag{28}$$

Its derivative is:

$$\Phi'_{sing\_anal}(z) = \frac{1}{2}\left(\frac{z}{\sqrt{z^2-1}} - 1\right). \tag{29}$$

Therefore, Taylor's expansion for an analytical solution can be presented as:

$$\Phi_{sing\_anal}(z) \approx \frac{1}{2}\frac{\zeta(z-\zeta)}{\sqrt{\zeta^2-1}} + \frac{1}{2}\sqrt{\zeta^2-1} - \frac{z}{2}, \quad \zeta \in \Gamma, \tag{30}$$

Numerical calculations of $\Phi_{sing}(\zeta)$ at the collocation points $\eta_m \ (m=1...n-1)$ produce:

$$\Phi_{sing\_num}(\eta_m) = \frac{1}{2\pi}\int_{-1}^{1}\frac{\sqrt{1-t^2}}{t-\zeta}dt = \frac{1-\eta_m^2}{2n}\sum_{k=1}^{n}\frac{1}{\xi_k - \eta_m} - \frac{\eta_m}{2},$$

$$\Phi_{sing\_num}(\eta_m) = \frac{1}{2\pi}\int_{-1}^{1}\frac{t}{\sqrt{1-t^2}(t-\eta_m)}dt = \frac{-1}{2n}\sum_{k=1}^{n}\frac{\xi_k}{\xi_k - \eta_m}. \tag{31}$$

For simplicity, let us assume that $\zeta = \eta_m + i\varepsilon$, where $\varepsilon \ll 1$. Thus, eqn (26) assumes the form:

$$\Phi_{\text{sing\_num}}(\zeta) = \frac{1-\eta_m^2}{2n} \sum_{k=1}^{n} \frac{1}{\xi_k - \eta_m} - \frac{\eta_m}{2} + i\frac{\sqrt{1-\eta_m^2}}{2} +$$

$$+ \left[\frac{1}{2n} \sum_{k=1}^{n} \frac{\xi_k}{\xi_k - \eta_m} + \frac{i}{2}\frac{\eta_m}{\sqrt{1-\eta_m^2}} \right] \varepsilon \qquad (32)$$

$$\Phi_{\text{sing\_num}}(\zeta) = \frac{1}{2n} \sum_{k=1}^{n} \frac{1-\eta_m^2 + \varepsilon\xi_k}{\xi_k - \eta_m} - \frac{\eta_m}{2} + \frac{i}{2}\frac{1-\eta_m^2 + \varepsilon\eta_m}{\sqrt{1-\eta_m^2}}.$$

In Table 1 we show the results of calculations using eqn (32) in comparison with the exact solution (28), analytical approximation (30) and calculations using the quadrature formula applied directly to eqn (27), for $n = 8$ and $\varepsilon = 0.5, 0.05$, and 0.005.

It is evident from Table 1 that analytical and numerical approximations (3rd and 4th columns) produce the same results, which can be explained by the high accuracy of the quadrature formula of the Chebyshev–Gauss type. The number of nodes $n = 8$ is even redundant, but it is used here to compare the results at more points. For $\varepsilon = 0.5$, the results using Taylor's approximations are unsatisfactory, while exact and direct calculations

Table 1: Exact, analytical, numerical and direct calculations of the complex potential.

Point coordinates	Exact (28)	Analytic approximation (30)	Numerical approximation (32)	Direct calculations of (27)
$\varepsilon = 0.5$				
$0.924 + 0.5$ i	$-0.186 + 0.169$ i	$-0.212 + 0.795$ i	$-0.212 + 0.795$ i	$-0.186 + 0.169$ i
$0.707 + 0.5$ i	$-0.166 + 0.222$ i	$-0.104 + 0.604$ i	$-0.104 + 0.604$ i	$-0.166 + 0.222$ i
$0.383 + 0.5$ i	$-0.102 + 0.283$ i	$0.059 + 0.565$ i	$0.059 + 0.565$ i	$-0.102 + 0.283$ i
0.5 i	0.309 i	$0.25 + 0.5$ i	$0.25 + 0.5$ i	0.309 i
$-0.383 + 0.5$ i	$0.102 + 0.283$ i	$0.441 + 0.358$ i	$0.441 + 0.358$ i	$0.102 + 0.283$ i
$-0.707 + 0.5$ i	$0.166 + 0.222$ i	$0.604 + 0.104$ i	$0.604 + 0.104$ i	$0.166 + 0.222$ i
$-0.924 + 0.5$ i	$0.186 + 0.169$ i	$0.712 - 0.412$ i	$0.712 - 0.412$ i	$0.186 + 0.169$ i
$\varepsilon = 0.05$				
$0.924 + 0.05$ i	$-0.405 + 0.176$ i	$-0.437 + 0.252$ i	$-0.437 + 0.252$ i	$-0.43 + 0.133$ i
$0.707 + 0.05$ i	$-0.329 + 0.33$ i	$-0.329 + 0.379$ i	$-0.329 + 0.379$ i	$-0.346 + 0.157$ i
$0.383 + 0.05$ i	$-0.181 + 0.438$ i	$-0.166 + 0.472$ i	$-0.166 + 0.472$ i	$-0.189 + 0.164$ i
0.05 i	0.476 i	$0.025 + 0.5$ i	$0.025 + 0.5$ i	0.165 i
$-0.383 + 0.05$ i	$0.181 + 0.438$ i	$0.216 + 0.452$ i	$0.216 + 0.452$ i	$0.189 + 0.164$ i
$-0.707 + 0.05$ i	$0.329 + 0.33$ i	$0.379 + 0.329$ i	$0.379 + 0.329$ i	$0.346 + 0.157$ i
$-0.924 + 0.05$ i	$0.405 + 0.176$ i	$0.487 + 0.131$ i	$0.487 + 0.131$ i	$0.43 + 0.133$ i
$\varepsilon = 0.005$				
$0.924 + 5E - 03$ i	$-0.456 + 0.189$ i	$-0.459 + 0.197$ i	$-0.459 + 0.197$ i	$-0.462 + 0.017$ i
$0.707 + 5E - 03$ i	$-0.351 + 0.351$ i	$-0.351 + 0.356$ i	$-0.351 + 0.356$ i	$-0.353 + 0.017$ i
$0.383 + 5E - 03$ i	$-0.19 + 0.459$ i	$-0.189 + 0.463$ i	$-0.189 + 0.463$ i	$-0.191 + 0.017$ i
$5E - 03$ i	0.498 i	$2.5E - 03 + 0.5$ i	$2.5E - 03 + 0.5$ i	0.017 i
$-0.383 + 5E - 03$ i	$0.19 + 0.459$ i	$0.194 + 0.461$ i	$0.194 + 0.461$ i	$0.191 + 0.017$ i
$-0.707 + 5E - 03$ i	$0.351 + 0.351$ i	$0.356 + 0.351$ i	$0.356 + 0.351$ i	$0.353 + 0.017$ i
$-0.924 + 5E - 03$ i	$0.456 + 0.189$ i	$0.464 + 0.185$ i	$0.464 + 0.185$ i	$0.462 + 0.017$ i

coincide. For ε = 0.05, all results are more or less of the same quality as exact solutions. For ε = 0.005, calculations using direct Chebyshev integration of eqn (27) produce incorrect results, especially for the imaginary part of the complex potential Φ(z). In the meantime, Taylor's approximations (30) and (32) are remarkably close to calculations using the exact eqn (28).

3.3 Example: Principal stress directions near a rectilinear crack

We will use the results of the previous subsection to illustrate what are typical values of errors in the calculation of principal direction near a rectangular crack. For this purpose, let us select the derivative of the potential Φ(z) in the form similar to eqn (27):

$$\Phi'(z) = \Phi'_S(z) + \Phi'_R(z), \quad \Phi'_S(z) = \frac{1}{2i\pi}\int_{-1}^{1}\frac{\sqrt{1-t^2}}{t-z}\,dt, \tag{33}$$

Here, $\Phi'_R(z)$ is given by a regular integral, when z tends to the contour point ζ. We can still accept that the condition of single valuedness is satisfied, because the density of $\Phi(z)$ is odd and equal to $\frac{t}{2}\sqrt{1-t^2} + \arcsin t$.

For rectilinear boundaries, the second potential is $\Psi(z) = -z\Phi'(z)$; therefore, $D(z,\bar{z}) = (\bar{z}-z)\Phi'(z)$ and the principal directions are determined by the argument of $\Phi'(z)$ alone. Simplify by omitting the regular term in eqn (33), to find an exact expression for the principal directions:

$$\theta(z) = -\frac{1}{2}\arg\left(\frac{-1}{\pi}\int_{-1}^{1}\frac{\sqrt{1-t^2}}{t-z}\,dt\right) = -\frac{1}{2}\arg\left(z - \sqrt{z^2-1}\right), \quad \mathrm{Im}(z) > 0. \tag{34}$$

On the other hand, using Taylor's expansion for the function $\Phi'_S(z)$ one can find an approximate expression for the principal directions at points $\eta_m + i\varepsilon$:

$$\theta_{\mathrm{app}}(\eta_m + i\varepsilon) \approx \frac{-1}{2}\arg\left(\frac{\eta_m^2 - 1 + i\varepsilon\eta_m}{\sqrt{\eta_m^2 - 1}} - \eta_m - i\varepsilon\right), \quad m = 1\ldots n-1. \tag{35}$$

We can further use this equation only without referencing numerical calculations, because the results of both numerical and analytical calculations fully coincide, as was evident from the 3rd and 4th columns of Table 1.

The results of calculations are summarised in Table 2 for ε = 0.5, 0.05, and 0.005. It is evident from Table 2 that approximation (35) is incorrect for ε = 0.5, while exact and direct calculations coincide. For ε = 0.05, both approximate and direct calculations possess reasonable accuracy. For ε = 0.005, the direct numerical calculations failed, while the approximation coincides with the exact calculations.

Table 2: Comparisons of exact, approximate and direct calculations of the principal directions (degrees).

Point coordinates	Exact (34)	Approximation (35)	Direct calculations from eqn (33)
$\varepsilon = 0.5$			
$0.924 + 0.5\ i$	21	38	21
$0.707 + 0.5\ i$	27	40	27
$0.383 + 0.5\ i$	35	48	35
$0.5\ i$	45	58	45
$-0.383 + 0.5\ i$	55	70	55
$-0.707 + 0.5\ i$	63	85	63
$-0.924 + 0.5\ i$	69	-75	69
$\varepsilon = 0.05$			
$0.924 + 0.05\ i$	12	15	9
$0.707 + 0.05\ i$	23	25	12
$0.383 + 0.05\ i$	34	35	20
$0.05\ i$	45	46	45
$-0.383 + 0.05\ i$	56	58	70
$-0.707 + 0.05\ i$	67	70	78
$-0.924 + 0.05\ i$	78	82	81
$\varepsilon = 0.005$			
$0.924 + 5E-3\ i$	11	12	1
$0.707 + 5\ E-3\ i$	23	23	1
$0.383 + 5\ E-3\ i$	34	34	3
$5\ E-3\ i$	45	45	45
$-0.383 + 5\ E-3\ i$	56	56	87
$-0.707 + 5\ E-3\ i$	67	68	89
$-0.924 + 5\ E-3\ i$	79	79	89

4 CONCLUSIONS

In this study, we investigated the applicability of Taylor's expansions and suggested an algorithm for the calculations of some characteristics of the stress state near the contour. We show that for small distances from the contour, direct application of quadrature formulas fails, while our proposed numerical approach produces highly accurate results.

ACKNOWLEDGEMENT

This work was partly supported by a grant from the Russian Foundation for Basic Research: grant number 20-05-00629 A.

REFERENCES

[1] Muskhelishvili, N.I., *Some Basic Problems of the Mathematical Theory of Elasticity*, P. Noordhoff: Groningen, the Netherlands, 1963.
[2] Savruk, M.P., *Two-Dimensional Problems of Elasticity for Body with Cracks*, Naukova Dumka: Kiev, Ukraine, 1981.
[3] Linkov, M., *Boundary Integral Equations in Elasticity Theory*, Kluwer, 2002.

[4]   Galybin, A.N. & Mukhamediev, Sh.A., Determination of elastic stresses from discrete data on stress orientations. *International Journal of Solids and Structures*, **41**(18–19), pp. 5125–5142, 2004.

[5]   Gakhov, F.D., *Boundary Value Problems*, Dover Publications: New York, 1990.

SECTION 2
COMPUTATIONAL METHODS

DEVELOPMENT OF A MESH CLUSTERING ALGORITHM AIMED AT REDUCING THE COMPUTATIONAL EFFORT OF GEARBOXES' CFD SIMULATIONS

MARCO NICOLA MASTRONE & FRANCO CONCLI
Free University of Bolzano, Italy

ABSTRACT

Over the last decade the development of simulation tools led to significant advancements in the design of virtual prototypes. Mechanical design benefitted a lot from this technological progress in terms of costs' reduction. However, in many applications as, for example, gearboxes, the implementation of complex CFD models that involve multiphase simulations with dynamic meshes and turbulence modelling is still a concern in terms of computational time. The main issue is related to the management of the topological modifications of the computational domain during the boundary rotation. For this reason, it is fundamental to have at disposal accurate and time efficient solutions for the correct evaluation of the efficiency and lubrication of geared transmissions already in the early stages of the design. In this work, an automatic algorithm for the reduction of the computational effort associated with the mesh generation and handling in the simulation of gearboxes is presented. The algorithm is based on the creation of a limited number of meshes for the description of the whole wheels' rotation by proper setting the time libraries taking advantage of the cyclicality of the gears' teeth position. Several operating conditions were simulated exploiting this methodology showing a drastic reduction of the simulations' time, making this tool ideal in the investigation of gears' lubrication and efficiency in a limited amount of time. The methodology was implemented in the open-source software OpenFOAM®.

Keywords: CFD, mesh handling, lubrication, efficiency, power losses, OpenFOAM®, gears.

1 INTRODUCTION

The technological advancements in computer science that have characterized the last decade offered new possibilities to approach mechanical design. Simulation codes are nowadays widespread both in industry and academia. The reduction of time to market enabled by virtual prototypes as well as the opportunity to investigate operating conditions for which data acquisition can be complex and costly are the fundamental reasons behind such diffusion. These tools are referred to as Computer-Aided Engineering (CAE) among which the most used are Finite Element Analysis (FEA), Multibody Dynamics (MBD) and Computational Fluid Dynamics (CFD). The application of these tools allowed engineers to support experimental findings with numerical results and to analyse (numerically) the behavior of the considered system in several operating conditions. One of the aspects that has benefitted a lot from simulation software packages is related to the study of lubrication and efficiency of geared transmissions. In fact, the application of numerical codes to gearboxes allows one to obtain information on the lubricant distribution and the power dissipation that are usually hard to quantify analytically or to measure experimentally (difficult optical access, need for niche equipment and simplified prototypes, etc.). For these reasons, CFD studies can be beneficial to bridge the gap leading to a deeper understanding of the physical phenomena and to tackle the more and more stringent requirements of energy efficient solutions. One of the main limitations in the simulations of gears is related to the high computational resources needed to handle the topological changes of the domain. In fact, while the gears rotate, the mesh loses its validity and must be updated with a new grid having sufficient quality to ensure the numerical convergence. It is common to need hundreds of meshes to accomplish the

WIT Transactions on Engineering Sciences, Vol 131, © 2021 WIT Press
www.witpress.com, ISSN 1743-3533 (on-line)
doi:10.2495/BE440051

gears' rotation and to reach the solution convergence. The massive need of valid grids has a severe impact on the computational time.

While it is already possible to efficiently simulate spur gears exploiting extrusion algorithms which reduce the problem to a 2–2.5D remeshing process [1]–[11], more complex geometries, as helical and bevel gears, cannot benefit from this solution, and tetrahedral meshes must be employed. These geometries have been simulated mainly by exploiting commercial codes requiring massive parallelization and high computational resources both with mesh [12]–[17] and meshless methods [18]–[20], while no work was found in the literature based on opensource software. The authors maintain that a general algorithm applicable for any system configuration can represent an important step in the simulation of gearboxes and in the reduction of computational resources needed for the meshes handling of the domain. In this work, an algorithm that requires a limited number of grids is implemented in the open-source software OpenFOAM® [21] showing a drastic reduction of the computational costs associated with the mesh generation opening new scenarios for a massive introduction of CFD applied to lubrication also in industry.

2 MATERIALS AND METHODS

2.1 Meshing strategy

To numerically discretize gearboxes, it is fundamental to use a robust mesh generator. Native applications of OpenFOAM® are *blockMesh* and *snappyHexMesh*. The former is a utility that builds full a structured mesh providing the user with a perfect control on every aspect of the mesh generation process. The main drawback of this tool is that it is not suitable for gearboxes due to the complex shape of the computational domain which would require very complex partitions. The latter is a utility that builds hexahedral dominant grids starting from STL files. However, it is time and memory consuming and does not represent the optimal solution for gearboxes simulation. Another alternative is represented by *cfMesh*, an external grid generator that, in its base version, can be installed freely as a library that communicates with OpenFOAM®. Even third-party software (commercial or opensource) can be exploited to create a mesh that can be converted into the OpenFOAM® format. One of the most used one is Salome [22], a powerful open-source pre-processor that can build polyhedral grids and has a Graphical User Interface (GUI) that makes the meshing process more intuitive. Since the correct setup of the CFD simulation of gearboxes is already demanding from a modeling point of view (aspects as multiphase conditions, dynamic mesh and turbulence must be considered simultaneously), it is fundamental to choose the meshing strategy that mostly reduces the complexity of the virtual model. In fact, the boundary rotation causes the distortion of the mesh which should be replaced with a new valid one at every little angular position. For this reason, a new algorithm called Global Remeshing Approach with Mesh Clustering (GRA^{MC}) was implemented. This strategy is based on the creation of a limited number of numerical meshes that covers one engagement. Indeed, after the first engagement, the wheels find themselves in the exact same position as the one of the first mesh, thus meaning that the remeshing process can be reduced to the creation of predefined grids that, in turn, can be recursively reused to describe the whole rotation of the gears. By doing so, the amount of meshes needed is dramatically reduced to just few grids, leading to very high computational performance. Despite being simple considerations, the correct implementation of such algorithm requires a specific procedure based not only on the mesh quality indicators, but also on the design parameters of the gearbox. Moreover, all the libraries that control the time must be updated automatically at each mesh substitution.

The first step is the creation of a high-quality mesh. In the authors experience, a maximum non-orthogonality of 70 and a maximum skewness of 2 are the threshold that should not be exceeded in these simulations. To find the number of meshes necessary to cover one engagement, a generic rotational speed is imposed to the gears. When the mesh quality exceeds the imposed limits, the simulation is stopped and the last valid timestep is taken as a reference ($t_{up_{guess}}$). The angle (α) between two successive teeth is calculated for each gear and the time needed for completing 1 engagement (T_{eng}) can be obtained:

$$\alpha = \frac{2\pi}{N_{teeth}}. \tag{1}$$

$$T_{eng} = \frac{\alpha}{\omega}. \tag{2}$$

The following step is to identify the updating time of each mesh (t_{up}) that allows T_{eng} to be reached with a set of predefined meshes. Exploiting the time $t_{up_{guess}}$ previously found, the number of meshes needed to cover 1 engagement (N_{mesh}) is then calculated as:

$$N_{mesh} = \frac{T_{eng}}{t_{up_{guess}}}. \tag{3}$$

N_{mesh} is then approximated to the nearest bigger integer number, and the $t_{up_{guess}}$ is changed accordingly to the actual time t_{up} that will be used to update the grids through the engagement.

Once the mesh set with the wheels in predefined positions has been computed, it can be used for the entire simulation. The governing equations are solved subsequently for each grid, the results are saved and mapped onto the following mesh that is immediately provided as input. The results' interpolation follows a *consistent* strategy that guarantees that all variables are mapped between conformal domains. Furthermore, thanks to the very good control over the mesh generation parameters, the interpolation occurs between very similar grids in terms of elements' size and, therefore, the numerical errors are significantly reduced. At regime, the results are postprocessed. While the updating of the time libraries is performed at the end of each t_{up}, the choice of which mesh to use at every step is managed by an *if-else* statement inside the procedure loop itself. The general workflow of the solution algorithm, which is completely automated in a Bash [23] script, is summarized in Fig. 1.

In the analyzed case, the final set of mesh is made of 10 grids. The time needed to create the set can be quantified in about 100 minutes. Without this algorithm, 230 mesh substitutions with a computational cost of 2,300 minutes would be necessary to cover just 1 gear complete rotation. Considering that for such simulation hundreds of mesh substitutions are necessary to reach convergence, the enormous advantages of the implemented algorithm can be clearly understood. The implementation of this strategy requires some coding to automatize the whole procedure.

2.2 Numerical approach

CFD codes are based on the solution of governing equations: mass, momentum, and energy conservation. In this study, the problem was modelled as isothermal. Therefore, the energy equation was not included in the calculation. In this way, the solution is limited to the mass and momentum equations which can be written as:

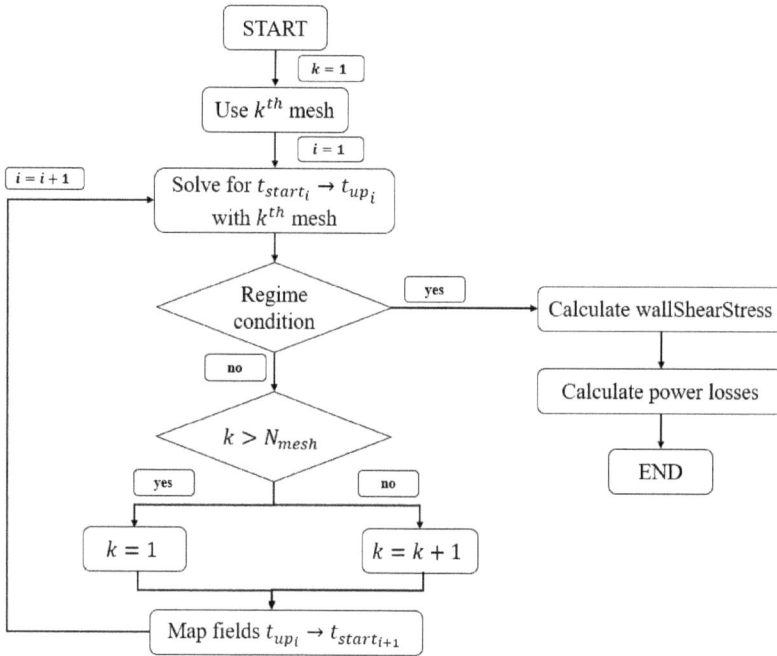

Figure 1: Scheme of the solution loop.

$$\frac{\partial \rho}{\partial t} + \nabla(\rho \boldsymbol{u}) = 0. \tag{4}$$

$$\frac{\partial(\rho \boldsymbol{u})}{\partial t} + \nabla(\rho \boldsymbol{u}\boldsymbol{u}) = -\nabla p + \nabla \left[\mu(\nabla \boldsymbol{u} + \nabla \boldsymbol{u}^T)\right] + \rho \boldsymbol{g} + \boldsymbol{F}. \tag{5}$$

These equations are valid only in simulations involving one phase. To model multiphase problems numerically, an additional balance equation to consider the presence of two or more phases must be added to the previous equation. By exploiting the volume of fluid (VOF) model [24], which is based on the definition of the scalar quantity volumetric fraction representing the percentage of one fluid in every cell of the domain, the multiphase problem can be solved. The equation of the volumetric fraction can be expressed as follows:

$$\frac{\partial \alpha}{\partial t} + \nabla(\alpha \boldsymbol{u}) = 0. \tag{6}$$

The properties Θ of the different fluids (such as density and viscosity) are taken to define the properties of an equivalent fluid as follows:

$$\Theta = \Theta \alpha + \Theta(1 - \alpha), \tag{7}$$

where Θ represents the generic property of each fluid.

The MULES (Multidimensional Universal Limiter with Explicit Solution) [25] correction can be added in the solver algorithm in order to obtain a more stable and bounded solution of the volumetric fraction field. This is accomplished by adding a dummy velocity field ($\boldsymbol{u_c}$) in the conservation equation of the volumetric fraction α:

$$\frac{\partial \alpha}{\partial t} + \nabla(\alpha \boldsymbol{u}) + \nabla(\boldsymbol{u}_c \alpha(1 - \alpha)) = 0. \tag{8}$$

An additional source term (S_g) must be added to the equation to account for additional phenomena as cavitation:

$$\frac{\partial \alpha}{\partial t} + \nabla(\alpha \boldsymbol{u}) + \nabla(\boldsymbol{u}_c \alpha(1 - \alpha))) = S_g. \tag{9}$$

To calculate the source term, a mathematical model must be introduced. The most used ones for describing cavitation are those by Kunz [26], Merkle et al. [27] and Saurer [28]. In this work, the Kunz model was used to model one operating condition (to account for cavitation, as will be explained in detail later). The great advantage of this formulation is related to the source term being independent of pressure. The vaporization ($\dot{m}_v$) is modelled as proportional to the liquid fraction (α) and to the quantity of pressure under the saturation pressure (p_{sat}). The condensation ($\dot{m}_c$) is modelled analogously:

$$\dot{m}_v = \frac{C_v \rho_v}{\frac{1}{2}\rho_l U_\infty^2 t_\infty} \alpha \min[0, p - p_{sat}]. \tag{10}$$

$$\dot{m}_c = \frac{C_c}{\frac{1}{2}U_\infty^2 t_\infty} (1 - \alpha) \max[0, p - p_{sat}]. \tag{11}$$

2.3 Analyzed system

A helical gearbox is considered in the current analysis. The geometrical parameters are reported in Table 1.

Table 1: Gears' geometrical characteristics.

	Unit	Gears
Transmission ratio	–	1
Number of teeth	–	23
Module	mm	4
Helix angle	°	27
Face width	mm	40

A local size of 1 mm was imposed on the wheels' edges for refinement purposes, whereas a global size of 5 mm for the rest of the domain was used. A growth rate defining the difference between two adjacent elements of 20% was imposed. Five surface and volume optimization loops were utilized to improve the mesh quality indicators. These settings allowed one to obtain high quality meshes that even at the maximum deformation (just before the new valid mesh is provided) have a maximum non-orthogonality and a maximum skewness below 70 and 2, respectively. The total number of cells is about 520k.

In Table 2 the simulated operating conditions are reported.

2.4 Numerical schemes

The PIMPLE (merged PISO-SIMPLE) algorithm was used in all simulations. This algorithm allows a better control in transient simulations. In fact, it is possible to tune the correctors of

Table 2: Operating conditions.

Operating condition	Rotational speed (rpm)	Oil density (kg/m³)	Oil viscosity (mm²/s)
Multiphase (centerline)	3,000		
	6,000	800	0.001
Complete filling	3,000		
	6,000		
Complete filling with cavitation	3,000		
	6,000		

the conservation equations within one timestep to reach the best compromise between computational effort and stability of the simulation. A convergence criterion of $1e-6$ was imposed to all field's variables. The GAMG (Generalized Geometric-Algebraic Multi-Grid) solver was used for the pressure, while the velocity was solved with the Gauss–Seidel smooth solver. A maximum Courant number of 1 was imposed to ensure the stability of the simulations. The first-order implicit Euler scheme was used for time-stepping. A Total Variation Diminishing (TVD) scheme using the vanLeer limiter was adopted for the convection of the volumetric fraction.

3 RESULTS

Fig. 2 shows the lubricant distribution in the multiphase condition. The lubricant remains entrapped between the teeth and tends to move along it due to the helical configuration. Pressure peaks arise on the teeth flanks in contact with lubricant due to the higher inertial effects.

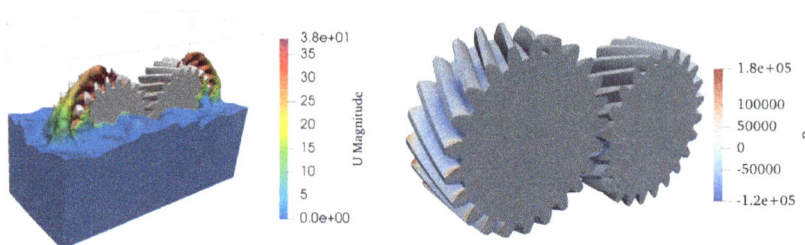

Figure 2: Lubricant distribution and pressure contour on the gears.

In Fig. 3 axial gradients in the meshing region are shown for the complete filled condition. These are related to the pressure increase in the meshing region. When the teeth leave the contact point, the pressure decreases, thus generating a suction effect which is highlighted by the axial velocity streamlines.

The simulation including cavitation shows that completely gearbox filling is not sufficient to ensure the complete wetting of the wheels. In fact, a pressurization is applied to accomplish this objective in practice [29]. If an external pressurization is not applied, local phase changes from liquid to vapour occur. In Fig. 4, the volumetric fraction field on a tooth flank is shown. Values lower than 1 indicate that the oil is cavitating.

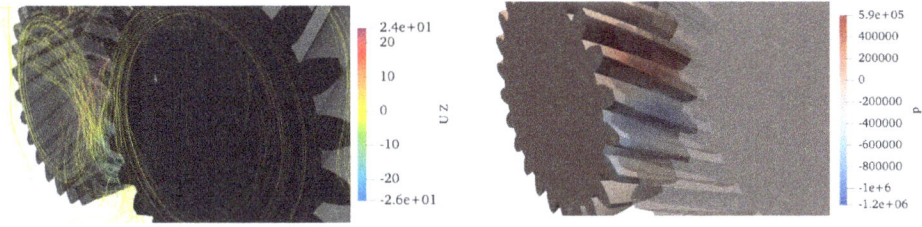

Figure 3: Axial velocity streamlines and pressure contour in the meshing zone.

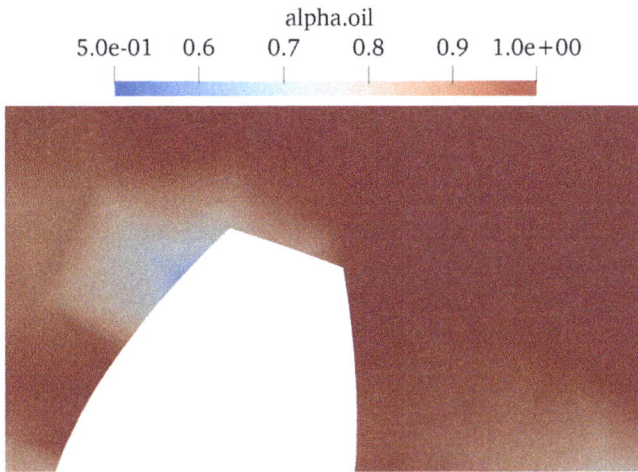

Figure 4: Lubricant contour on a slice of the gear in the cavitation case.

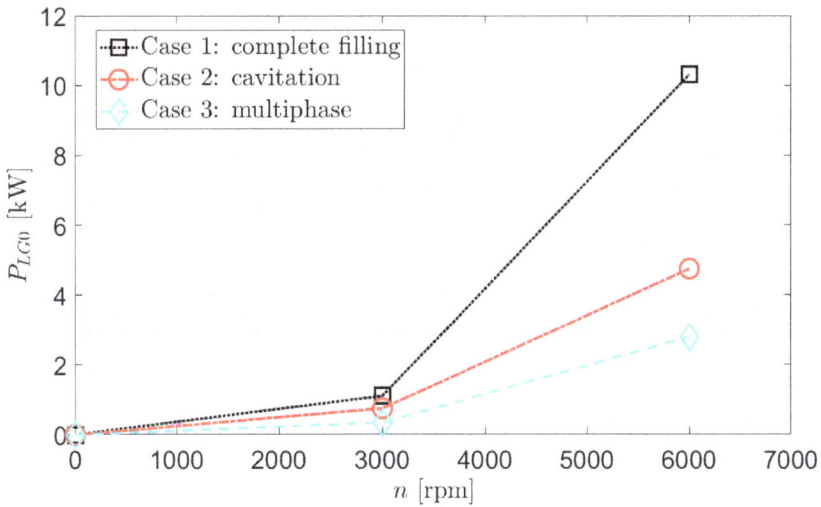

Figure 5: Power losses vs rotational speed for the different operating condition.

The power dissipations for the three configurations is reported in Fig. 5. It can be noticed that the complete filling condition is characterized by the highest losses. This is mainly related to the windage effects, while in the multiphase condition is the oil splashing responsible for the power losses.

4 COMPUTATIONAL PERFORMANCE

The simulations were performed on a Deploy LXD Compute Node (2xINTEL Xeon® E-2680, 8 Cores, 2.7 GHz). The simulations required on average 10 h each to reach convergence. The low computational time achieved parallelizing the computations among only 8 processors is mainly related to the implemented meshing algorithm that relies on the usage of just 10 meshes for the whole simulation. The computational resources dedicated to mesh management during the simulations are therefore drastically reduced and devoted just for the mapping process from mesh to mesh, since all the grids have been computed in advance. A comparison between the implemented GRA^{MC} and a standard remeshing strategy that does not exploit the cyclicality of the gears' teeth position is reported in Table 3.

Table 3: GRA^{MC} computational gain normalized to one gear rotation.

Remeshing algorithm	STD approach	GRA^{MC}	Net gain %
Parametric value (1 rotation of the wheel)	t	$\dfrac{t}{N_{teeth}}$	$\left(1 - \dfrac{1}{N_{teeth}}\right) \times 100$
Specific value for the analyzed gears	2,300 min	100 min	95.66%

Normalizing the computational effort to 1 gear rotation, the net gain associated with the explained mesh reduction algorithm can be quantified as 95.66%. This has positive impact on the simulation time itself which is drastically reduced. All the time libraries are defined to ensure the wheels' passage in the control positions defined by the computed mesh set.

In Fig. 6, the relative time gain vs the number of gears rotation is shown for gears with different number of teeth. As the number of rotations increases, the effects of the GRA^{MC}

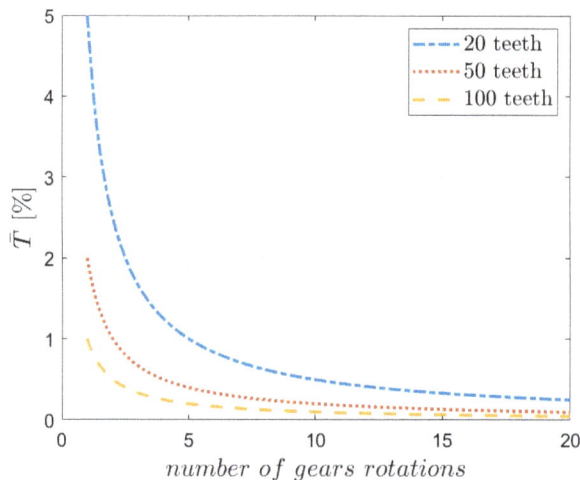

Figure 6: Meshing time for different gears' rotations and gears' size.

become even more evident since the simulation time reduces continuously. This indicates that this algorithm represents an effective strategy to reduce the computational time in the CFD simulation of various gear types.

5 CONCLUSIONS

Differently from previous analyses, a complex domain as a helical gearbox was simulated in an opensource environment with a general procedure applicable to any configuration. The algorithm is based on the calculation of a set of grids that covers one engagement. This set is recursively reused at each engagement, thus limiting the complexity of the remeshing process. The effectiveness of the implemented procedure allowed a drastic reduction of the simulation time associated with the mesh management, which is one of the major concerns in these models. While the aim of this study was to show the capabilities of the new approach and the reduction of mesh requirements, from the simulations some interesting considerations emerged regarding the physics involved in lubricated systems and the relative differences in various operating conditions. Considering the complexity of obtaining such results experimentally, this methodology represents a time efficient solution even on a medium hardware without the need for big clusters.

REFERENCES

[1] Liu, H., Jurkschat, T., Lohner, T. & Stahl, K., Determination of oil distribution and churning power loss of gearboxes by finite volume CFD method. *Tribology International*, **109**, pp. 346–354, 2017. DOI: 10.1016/j.triboint.2016.12.042.

[2] Mastrone, M.N., Hartono, E.A., Chernoray, V. & Concli, F., Oil distribution and churning losses of gearboxes: Experimental and numerical analysis. *Tribology International*, **151**, 2020. DOI: 10.1016/j.triboint.2020.106496.

[3] Concli, F. et al., Load independent power losses of ordinary gears: Numerical and experimental analysis. *5th World Tribology Congress, WTC 2013*, vol. 2, pp. 1243–1246, 2013. https://www.scopus.com/inward/record.uri?eid=2-s2.0-84919653123&partnerID=40&md5=1df4c1daa085b17da52935814f251c5c.

[4] Concli, F. & Gorla, C., CFD simulation of power losses and lubricant flows in gearboxes. *American Gear Manufacturers Association Fall Technical Meeting 2017*, 2017.

[5] Concli, F. & Gorla, C., Influence of lubricant temperature, lubricant level and rotational speed on the churning power loss in an industrial planetary speed reducer: Computational and experimental study. *International Journal of Computional Methods & Experimental Measurements*, **1**(4), pp. 353–366, 2013. DOI: 10.2495/CMEM-V1-N4-353-366.

[6] Concli, F. & Gorla, C., A CFD analysis of the oil squeezing power losses of a gear pair. *International Journal of Computional Methods & Experimental Measurements*, **2**(2), pp. 157–167, 2014. DOI: 10.2495/CMEM-V2-N2-157-167.

[7] Concli, F., Della Torre, A., Gorla, C. & Montenegro, G., A new integrated approach for the prediction of the load independent power losses of gears: Development of a mesh-handling algorithm to reduce the CFD simulation time. *Advances in Tribology*, **2016**, pp. 1–8, 2016. DOI: 10.1155/2016/2957151.

[8] Concli, F. & Gorla, C., Influence of lubricant temperature, lubricant level and rotational speed on the churning power loss in an industrial planetary speed reducer: Computational and experimental study. *International Journal of Computional Methods & Experimental Measurements*, **1**(4), pp. 353–366, 2013.

[9]   Burberi, E., Fondelli, T., Andreini, A., Facchini, B. & Cipolla, L., CFD simulations of a meshing gear pair. *Proceedings of the ASME Turbo Expo*, vol. 5A-2016, 2016. DOI: 10.1115/GT2016-57454.

[10]   Concli, F., Low-loss gears precision planetary gearboxes: Reduction of the load dependent power losses and efficiency estimation through a hybrid analytical-numerical optimization tool [Hochleistungs- und Präzisions-Planetengetriebe: Effizienzschätzung und Reduzierun]. *Forschung im Ingenieurwesen/Engineering Research*, **81**(4), pp. 395–407, 2017. DOI: 10.1007/s10010-017-0242-0.

[11]   Bianchini, C., Da Soghe, R., Errico, J.D. & Tarchi, L., Computational analysis of windage losses in an epicyclic gear train. *Proceedings of the ASME Turbo Expo*, vol. 5B-2017, 2017. DOI: 10.1115/GT2017-64917.

[12]   Dai, Y., Ma, F., Zhu, X., Su, Q. & Hu, X., Evaluation and optimization of the oil jet lubrication performance for orthogonal face gear drive: Modelling, simulation and experimental validation. *Energies*, **12**(10), 2019. DOI: 10.3390/en12101935.

[13]   Ferrari, C. & Marani, P., Study of air inclusion in lubrication system of CVT gearbox transmission with biphasic CFD simulation. *BATH/ASME 2016 Symposium on Fluid Power & Motion Control*, 2016. DOI: 10.1115/FPMC2016-1767.

[14]   Hu, X., Jiang, Y., Luo, C., Feng, L. & Dai, Y., Churning power losses of a gearbox with spiral bevel geared transmission. *Tribology International*, **129**, pp. 398–406, 2019. DOI: 10.1016/j.triboint.2018.08.041.

[15]   Peng, Q., Zhou, C., Gui, L. & Fan, Z., Investigation of the lubrication system in a vehicle axle: Numerical model and experimental validation. *Proceedings of the IMechE Part D: Journal of Automobile Engineering*, **233**(5), pp. 1232–1244, 2019. DOI: 10.1177/0954407018766128.

[16]   Peng, Q., Gui, L. & Fan, Z., Numerical and experimental investigation of splashing oil flow in a hypoid gearbox. *Engineering Applications of Computational Fluid Mechanics*, **12**(1), pp. 324–333, 2018. DOI: 10.1080/19942060.2018.1432506.

[17]   Dai, Y., Jia, J., Ouyang, B. & Bian, J., Determination of an optimal oil jet nozzle layout for helical gear lubrication: Mathematical modeling, numerical simulation, and experimental validation. *Complexity*, **2020**, 2020. DOI: 10.1155/2020/2187027.

[18]   Deng, X. et al., Lubrication mechanism in gearbox of high-speed railway trains. *Journal of Advanced Mechanical Design, Systems & Manufacturing*, **14**(4), 2020. DOI: 10.1299/jamdsm.2020jamdsm0054.

[19]   Deng, X., Wang, S., Hammi, Y., Qian, L. & Liu, Y., A combined experimental and computational study of lubrication mechanism of high precision reducer adopting a worm gear drive with complicated space surface contact. *Tribology International*, **146**, 2020. DOI: 10.1016/j.triboint.2020.106261.

[20]   Morhard, B., Schweigert, D., Mileti, M., Sedlmair, M., Lohner, T. & Stahl, K., Efficient lubrication of a high-speed electromechanical powertrain with holistic thermal management. *Forschung im Ingenieurwesen/Engineering Research*, 2020. DOI: 10.1007/s10010-020-00423-0.

[21]   OpenFOAM. http://www.openfoam.com.

[22]   Salome. http://www.salome-platform.org.

[23]   Bash. www.gnu.org/software/bash.

[24]   Hirt, C.W. & Nichols, B.D., Volume of fluid (VOF) method for the dynamics of free boundaries. *Journal of Computational Physics*, **39**(1), pp. 201–225, 1981.

[25]   Rusche, H., *Computational Fluid Dynamics of Dispersed Two-Phase Flows at High Phase Fractions*, Imperial College of Science, Technology and Medicine: London, 2002.

[26] Kunz, R.F. et al., Preconditioned Navier–Stokes method for two-phase flows with application to cavitation prediction. *Computers & Fluids*, **29**(8), pp. 849–875, 2000.

[27] Merkle, C.L., Feng, J. & Buelow, P.E.O., Computational modeling of the dynamics of sheet cavitation. *3rd International Symposium on Cavitation*, pp. 47–54, 1998.

[28] Saurer, J., *Instationären kaviterende Sträömung – Ein neues Modell, basierend auf Front Capturing (VoF) and Blasendynamik*, Universität Karlsruhe, 2000.

[29] Höhn, B.R., Michaelis, K. & Otto, H.P., Influences on no-load gear losses. *3rd European Conference on Tribology*, pp. 639–644, 2011.

LOCALIZED COLLOCATION TREFFTZ METHOD IN CONJUNCTION WITH THE GENERALIZED RECIPROCITY METHOD FOR TRANSIENT HEAT CONDUCTION ANALYSIS IN HETEROGENEOUS MATERIALS

ZHUOJIA FU[1,2], MIN YANG[1], QIANG XI[1] & WENZHI XU[1]
[1]Center for Numerical Simulation Software in Engineering and Sciences,
College of Mechanics and Materials, Hohai University, China
[2]State Key Laboratory of Mechanics and Control of Mechanical Structures,
Nanjing University of Aeronautics & Astronautics, China

ABSTRACT

This paper presents a novel localized collocation Trefftz method (LCTM) in conjunction with Laplace transformation for transient heat conduction analysis in heterogeneous materials under temperature loading. In contrast to the conventional CTM, the proposed LCTM divides the whole domain into many stencil support domains consisting of several discretization nodes. Inspired by the dual reciprocity method (DRM) and multiple reciprocity method (MRM), an efficient technique, the generalized reciprocity method (GRM), is proposed to derive the problem-dependent T-complete functions for approximating the particular solution of the nonhomogeneous heat conduction equations in the local subdomains. Based on the moving least square technique and T-complete functions, the LCTM numerical differentiation formulation at a certain node can be derived by using a linear combination of the T-complete functions at its adjacent discretization nodes in the related stencil support domain. It inherits the semi-analytical property from the conventional CTM and avoids the ill-conditioned dense matrix problem, which is present particularly in large-scale heat conduction analysis. Some numerical examples of heat conduction problems in heterogeneous materials are presented, and the numerical results demonstrate the accuracy and efficiency of the proposed LCTM in comparison with the known analytical solutions.

Keywords: T-complete functions, collocation methods, Laplace transformation, heat conduction, moving least square, dual reciprocity method, multiple reciprocity method.

1 INTRODUCTION

With the increasing demand for long-term service of equipment structures in high-temperature environments, numerical simulation plays an important role on the prediction of the heat conduction behaviors of the equipment structures [1]–[3]. Here we consider the following transient heat conduction equations with thermal loading $Q(\mathbf{x},t)$ in heterogeneous materials,

$$\Re u(\mathbf{x},t) = Q(\mathbf{x},t), \qquad \mathbf{x} \in \Omega, t \in [0,T], \tag{1}$$

$$\mathrm{B}u(\mathbf{x},t) = g_B(\mathbf{x},t), \qquad \mathbf{x} \in \partial\Omega, t \in [0,T], \tag{2}$$

$$u(\mathbf{x},0) = g_I(\mathbf{x}), \qquad \mathbf{x} \in \Omega, \tag{3}$$

WIT Transactions on Engineering Sciences, Vol 131, © 2021 WIT Press
www.witpress.com, ISSN 1743-3533 (on-line)
doi:10.2495/BE440061

where $u(\mathbf{x},t)$ denotes the temperature, $g_B(\mathbf{x},t)$, $g_I(\mathbf{x})$ stands for the prescribed boundary condition functions and initial condition functions, the differential operators $\Re=\sum_{i,j=1}^{2}\dfrac{\partial}{\partial x_i}\left(K_{ij}\dfrac{\partial}{\partial x_j}\right)-\rho c\dfrac{\partial}{\partial t}$ and $\mathbf{B}=p_1+p_2\sum_{i,j=1}^{2}K_{ij}\dfrac{\partial}{\partial x_j}n_i$, in which

$$\mathbf{B}=\begin{cases}\textit{Dirichlet (essential) boundary condition operator, when } p_1=1, p_2=0,\\ \textit{Neumann (natural) boundary condition operator, when } p_1=0, p_2=1,\\ \quad\textit{Convective boundary condition operator, when } p_1\neq 0, p_2\neq 0,\end{cases}\quad\text{and}$$

$\mathbf{K}=\{K_{ij}\}_{1\leq i,j\leq 2}$ represents the thermal conductivity matrix, ρ stands for the density, c stands for the specific heat, n_i denotes the components of the outward unit normal vector in the x_i-direction.

2 METHODOLOGY

This section introduces the numerical method used here to solve eqns (1)–(3), which includes Laplace transformation, localized collocation Trefftz method [4], [5], generalized reciprocity method and numerical inverse Laplace transformation [6]–[9].

2.1 Laplace transformation

Here Laplace transformation is used to convert eqns (1)–(3) to time-independent PDEs in Laplace-space domain. Let us set the definition of Laplace transformation as

$$\mathfrak{I}^{(L)}(u(\mathbf{x},t))=u^{(L)}(\mathbf{x},p)=\int_{0}^{\infty}u(\mathbf{x},t)e^{-pt}dt, \tag{4}$$

then we get

$$\Re^{(L)}u^{(L)}(\mathbf{x},p)=Q_g^{(L)}(\mathbf{x},p),\quad \mathbf{x}\in\Omega, \tag{5}$$

$$\mathbf{B}u^{(L)}(\mathbf{x},p)=g_B^{(L)}(\mathbf{x},p),\quad \mathbf{x}\in\partial\Omega, \tag{6}$$

where the transformed governing equation operator $\Re^{(L)}=\sum_{i,j=1}^{2}\dfrac{\partial}{\partial x_i}\left(K_{ij}\dfrac{\partial}{\partial x_j}\right)-\rho cp$ and the generalized source function $Q_g^{(L)}(\mathbf{x},p)=-Q^{(L)}(\mathbf{x},p)-\rho cg_I(\mathbf{x})$, p stands for the Laplace transformation parameter, and the physical quantities in Laplace-space domain are represented by the superscript "(L)".

2.2 Localized collocation solver

This study employs a novel localized collocation solver for solving time-independent PDEs (5)–(6) in Laplace-space domain. In the localized collocation solver, the localized collocation Trefftz method (LCTM) in conjunction with generalized reciprocity method (GRM) is implemented, where the generalized reciprocity method (GRM) is inspired from the dual reciprocity method (DRM) [10] and multiple reciprocity method (MRM) [11] to approximate the particular solution of the nonhomogeneous PDEs in the local subdomains. When the T-complete function of the transformed governing equation operator

$\Re^{(L)} = \sum_{i,j=1}^{2} \frac{\partial}{\partial x_i} \left(K_{ij} \frac{\partial}{\partial x_j} \right) - \rho c p$ at each discretization node can be obtained, the LCTM + GRM

can be used directly. Otherwise, the analogy differential operator $\tilde{\Re}^{(L)}$ to $\Re^{(L)}$ should be determined according to the easy derivation of the related T-complete functions. Then eqn (5) can be rewritten as

$$\tilde{\Re}^{(L)} u^{(L)} (\mathbf{x}, p) = \tilde{Q}_g^{(L)} (\mathbf{x}, p), \quad \mathbf{x} \in \Omega, \tag{7}$$

where $\tilde{Q}_g^{(L)} (\mathbf{x}, p) = Q_g^{(L)} (\mathbf{x}, p) + \left(\tilde{\Re}^{(L)} - \Re^{(L)} \right) u^{(L)} (\mathbf{x}, p)$.

By adopting Atkinson's splitting approach, the approximate solution of eqns (5)/(7) and (6) can be expressed as

$$u^{(L)} (\mathbf{x}, p) = u_h^{(L)} (\mathbf{x}, p) + u_p^{(L)} (\mathbf{x}, p), \tag{8}$$

where $u_h^{(L)} (\mathbf{x}, p), u_p^{(L)} (\mathbf{x}, p)$ stands for the homogeneous and the particular solutions, respectively. Assuming that the particular solution $u_p^{(L)} (\mathbf{x}, p)$ satisfies

$$\Re^{(L)} u_p^{(L)} (\mathbf{x}, p) = Q_g^{(L)} (\mathbf{x}, p) \text{ or } \tilde{\Re}^{(L)} u_p^{(L)} (\mathbf{x}, p) = \tilde{Q}_g^{(L)} (\mathbf{x}, p), \tag{9}$$

and then the homogeneous solution can be obtained by solving the following updated homogeneous equation

$$\Re^{(L)} u_h^{(L)} (\mathbf{x}, p) = 0 \quad \text{or} \quad \tilde{\Re}^{(L)} u_h^{(L)} (\mathbf{x}, p) = 0, \tag{10}$$

subjected to the updated boundary conditions

$$B u_h^{(L)} (\mathbf{x}, p) = g_B^{(L)} (\mathbf{x}, p) - B u_p^{(L)} (\mathbf{x}, p). \tag{11}$$

For each given i-th node $\mathbf{x}_0^i$, the related set of its m nearest nodes $\left(\mathbf{x}_1^i, \mathbf{x}_2^i, \dots, \mathbf{x}_m^i \right)$ around $\mathbf{x}_0^i$ can be found and named as a subdomain Ξ_i, and the center of this subdomain can be set as $\tilde{\mathbf{x}}^i = \frac{1}{m+1} \sum_{j=0}^{m} \mathbf{x}_j^i$. Then the generalized reciprocity method (GRM) introduces the associated differential operator $\Re_1^{(L)}$ to vanish the generalized source term $Q_g^{(L)} \left(\tilde{Q}_g^{(L)} \right)$ at each discretization node, namely,

$$\Re_1^{(L)} \Re^{(L)} u_p^{(L)} (\mathbf{x}, p) = \Re_1^{(L)} Q_g^{(L)} (\mathbf{x}, p) = \left(\Delta - \frac{\Delta Q_g^{(L)} (\mathbf{x}, p)}{Q_g^{(L)} (\mathbf{x}, p)} \right) Q_g^{(L)} (\mathbf{x}, p) = 0 \text{ or}$$

$$\Re_1^{(L)} \tilde{\Re}^{(L)} u_p^{(L)} (\mathbf{x}, p) = \Re_1^{(L)} \tilde{Q}_g^{(L)} (\mathbf{x}, p) = \left(\Delta - \frac{\Delta \tilde{Q}_g^{(L)} (\mathbf{x}, p)}{\tilde{Q}_g^{(L)} (\mathbf{x}, p)} \right) \tilde{Q}_g^{(L)} (\mathbf{x}, p) = 0, \tag{12}$$

where $\mathfrak{R}_1^{(L)}$ could be Laplace-type, Helmholtz-type and Modified-Helmholtz-type operators according to the value of $\dfrac{\Delta Q_g^{(L)}(\mathbf{x},p)}{Q_g^{(L)}(\mathbf{x},p)}\left(\dfrac{\Delta \tilde{Q}_g^{(L)}(\mathbf{x},p)}{\tilde{Q}_g^{(L)}(\mathbf{x},p)}\right)$.

Assuming that ϕ_k^i and φ_k^i stands for the derived T-complete functions of $\mathfrak{R}_1^{(L)}$ and $\mathfrak{R}^{(L)}\left(\tilde{\mathfrak{R}}^{(L)}\right)$, the approximated formulation $u^{(L)}\left(\mathbf{x}_j^i,p\right)$ inside the related subdomain Ξ_i can be represented by a linear combination of T-complete functions ϕ_k^i and φ_k^i with unknown coefficients α_k^i and β_k^i

$$u^{(L)}\left(\mathbf{x}_j^i,p\right)=\sum_{k=0}^{m}\phi_k^i\alpha_k^i+\sum_{k=0}^{m}\varphi_k^i\beta_k^i \text{ with its matrix form } \mathbf{u}^{(L)}=\begin{bmatrix}\mathbf{\Phi}^i & \mathbf{\Psi}^i\end{bmatrix}\begin{bmatrix}\mathbf{\alpha}^i\\\mathbf{\beta}^i\end{bmatrix}=\mathbf{\Theta}^i\mathbf{\chi}^i. \quad (13)$$

By employing the moving least square technique, the following function can be defined

$$\Lambda\left(u^{(L)}\left(\mathbf{x}_j^i,p\right)\right)=\sum_{j=1}^{N}\left[\left(u^{(L)}\left(\mathbf{x}_j^i,p\right)-\mathbf{\Theta}^i\mathbf{\chi}^i\right)w^i\left(d_j\right)\right]^2, \quad (14)$$

in which the weighting function $w^i\left(d_j\right)$ is defined as the following compact support quartic spline function given in the literature [12],

$$w^i\left(d_j\right)=\begin{cases}1-6\left(\dfrac{d_j}{d_{max}}\right)^2+8\left(\dfrac{d_j}{d_{max}}\right)^3-3\left(\dfrac{d_j}{d_{max}}\right)^4, & d_j\le d_{max} \\ 0, & d_j>d_{max}\end{cases}, \quad (15)$$

where d_{max} denote the maximum distance between central node $\tilde{x}^i=\dfrac{1}{m+1}\sum_{j=0}^{m}x_j^i$ and the nodes of its subdomain Ξ_i.

In order to get the LCTM approximated eqn (13), let us minimize the function $\Lambda\left(u^{(L)}\right)$ to determine the unknown coefficient $\mathbf{\chi}^i$ and then generate the following linear equation system at each discretization node,

$$\mathbf{D}\mathbf{\chi}^i=\mathbf{b}_R, \quad (16)$$

where

$$
\mathbf{D} =
\begin{bmatrix}
\sum\limits_{j=1}^{N}\Theta_{j1}^{2}w_{j}^{i} & \sum\limits_{j=1}^{N}\Theta_{j1}\Theta_{j2}w_{j}^{i} & \sum\limits_{j=1}^{N}\Theta_{j1}\Theta_{j3}w_{j}^{i} & \cdots & \sum\limits_{j=1}^{N}\Theta_{j1}\Theta_{jN}w_{j}^{i} \\[2ex]
 & \sum\limits_{j=1}^{N}\Theta_{j2}^{2}w_{j}^{i} & \sum\limits_{j=1}^{N}\Theta_{j2}\Theta_{j3}w_{j}^{i} & \cdots & \sum\limits_{j=1}^{N}\Theta_{j2}\Theta_{jN}w_{j}^{i} \\[2ex]
 & & \sum\limits_{j=1}^{N}\Theta_{j3}^{2}w_{j}^{i} & \cdots & \sum\limits_{j=1}^{N}\Theta_{j3}\Theta_{jN}w_{j}^{i} \\[2ex]
 & SYM & & \ddots & \vdots \\[2ex]
 & & & & \sum\limits_{j=1}^{N}\Theta_{jN}^{2}w_{j}^{i}
\end{bmatrix},
$$

$$
\mathbf{b}_{R} =
\begin{bmatrix}
\Theta_{11}w_{1}^{i} & \Theta_{21}w_{2}^{i} & \cdots & \Theta_{N1}w_{N}^{i} \\
\Theta_{12}w_{1}^{i} & \Theta_{22}w_{2}^{i} & \cdots & \Theta_{N2}w_{N}^{i} \\
\vdots & \vdots & \ddots & \vdots \\
\Theta_{1N}w_{1}^{i} & \Theta_{2N}w_{2}^{i} & \cdots & \Theta_{NN}w_{N}^{i}
\end{bmatrix}
\begin{bmatrix}
u^{(L)}\left(\mathbf{x}_{1}^{i},p\right) \\
u^{(L)}\left(\mathbf{x}_{2}^{i},p\right) \\
\vdots \\
u^{(L)}\left(\mathbf{x}_{N}^{i},p\right)
\end{bmatrix}
= \mathbf{b}_{D}\left(\mathbf{u}^{i}\right)^{(L)}.
$$

Assuming that the matrix $\mathbf{D}$ is reversible, one can get

$$
\boldsymbol{\chi}^{i} = \mathbf{D}^{-1}\mathbf{b}_{D}\left(\mathbf{u}^{i}\right)^{(L)} = \mathbf{W}\left(\mathbf{u}^{i}\right)^{(L)}. \tag{17}
$$

Next substituting eqn (17) into eqn (13), the Laplace-space solutions at $\left(\mathbf{x}_{1}^{i},p\right)$ can be represented as

$$
\left(u_{1}^{i}\right)^{(L)} = \boldsymbol{\Theta}_{1}^{i}\mathbf{D}^{-1}\mathbf{b}_{D}\left(\mathbf{u}^{i}\right)^{(L)} = \sum_{j=1}^{N}W_{j}^{i}\left(u_{j}^{i}\right)^{(L)}, \tag{18}
$$

where the weighting matrix $\mathbf{W}$ is a sparse matrix with ns nonzero elements in one row, i.e.
$\begin{cases} W_{j}^{i} \neq 0, \mathbf{x}_{j}^{i} \in \Xi_{i} \subset \Omega \\ W_{j}^{i} = 0, \mathbf{x}_{j}^{i} \in \Omega \backslash \Xi_{i} \end{cases}$, ns denotes the number of discretization nodes in the subdomain Ξ_{i}.

Then eqns (5), (7) and (6) can be discretized as follows:

$$
\mathfrak{R}^{(L)}\sum_{j=1}^{N}W_{j}^{i}u^{(L)}\left(\mathbf{x}_{j}^{i},p\right)=Q_{g}^{(L)}\left(\mathbf{x}_{j}^{i},p\right),
$$
$$
\text{or} \quad \tilde{\mathfrak{R}}^{(L)}\sum_{j=1}^{N}W_{j}^{i}u^{(L)}\left(\mathbf{x}_{j}^{i},p\right)=\tilde{Q}_{g}^{(L)}\left(\mathbf{x}_{j}^{i},p\right), \qquad \mathbf{x}_{j}^{i}\in\Omega, \tag{19}
$$

$$
\mathrm{B}\sum_{j=1}^{N}W_{j}^{i}u^{(L)}\left(\mathbf{x}_{j}^{i},p\right) = g_{B}^{(L)}\left(\mathbf{x}_{j}^{i},p\right), \quad \mathbf{x}_{j}^{i}\in\partial\Omega. \tag{20}
$$

Then the Laplace-space solutions $u^{(L)}\left(\mathbf{x}_{j}^{i},p\right)$ can be obtained by solving the system of linear algebra eqns (19) and (20).

2.3 Numerical inverse Laplace transformation

Here we introduce one of the well-established numerical inverse Laplace transformation algorithms, the fixed Talbot algorithm (FTA) [7], [8], to regain the time-dependent solution $u(\mathbf{x},t)$ in the time-domain from the Laplace-space solutions $u^{(L)}(\mathbf{x}^i_j, p)$. The related computational formulation of fixed Talbot algorithm can be given as follow

$$u(\mathbf{x}^i_j, T) = \frac{\hat{\xi}}{N_{LT}} \left\{ \begin{array}{l} \frac{1}{2} u^{(L)}(\mathbf{x}^i_j, \hat{\xi}) e^{\xi T} \\ + \sum_{j=1}^{N_{LT}-1} \mathrm{Re}\left[e^{T\omega(\eta_j)} u^{(L)}(\mathbf{x}, \hat{\omega}(\eta_j))\left(1 + \mathrm{i}\left(\hat{\zeta}(\eta_j)\right)\right) \right] \end{array} \right\}, \tag{21}$$

where

$$\hat{\omega}(\eta_j) = \hat{\xi}\eta_j\left(\cot\eta_j + \mathrm{i}\right), \quad \hat{\zeta}(\eta_j) = \eta_j + \left(\eta_j \cot\eta_j - 1\right)\cot\eta_j,$$

$$\hat{\xi} = \frac{2N_{LT}}{5T}, \quad \eta_j = \frac{j\pi}{N_{LT}}, \quad j = 1,2,\cdots,N_{LT} - 1.$$

For a specific time instant T, only N_{LT} boundary value problems with the corresponding Laplace-transform parameter $p = \hat{\xi}$ and $\omega(\eta_j)$ are required to be solved. In this study, $N_{LT} = 8$ is employed in the fixed Talbot algorithm.

3 NUMERICAL RESULTS AND DISCUSSIONS

This section presents the example to verify the efficiency of the proposed localized collocation Trefftz method (LCTM) in the solution of long-time heat conduction behaviour under heterogeneous materials. To measure the accuracy of the proposed LCTM, the relative error *Rerr* and L2 relative error *Lerr* are defined as follows,

$$\mathrm{Rerr}(u) = \left| \frac{u(\mathbf{x}^i, T) - u_{exact}(\mathbf{x}^i, T)}{u_{exact}(\mathbf{x}^i, T)} \right|, \tag{22}$$

$$\mathrm{Lerr}(u) = \sqrt{\frac{\sum_{i=1}^{N}\left(u(\mathbf{x}^i, T) - u_{exact}(\mathbf{x}^i, T)\right)^2}{\sum_{i=1}^{N} u_{exact}^2(\mathbf{x}^i, T)}}, \tag{23}$$

where $u(\mathbf{x}^i, T), u_{exact}(\mathbf{x}^i, T)$ stands for the numerical solutions and analytical solutions on the node $\mathbf{x}^i$ at the time instant T. Unless otherwise specified, the number of T-complete functions is chosen as $N_T = 11$ and the k-nearest neighbors algorithm is used to select $m = 12$ nearest nodes of the subdomain Ξ_i,

Here we consider the transient heat conduction problem in 2D square functionally graded material.

$$\nabla\left(K(\mathbf{x})\nabla u(\mathbf{x},t)\right) - \rho c \frac{\partial u(\mathbf{x},t)}{\partial t} = Q(\mathbf{x},t), \quad \mathbf{x} \in \Omega, t \in [0,T], \tag{24}$$

$$Bu(x_1, x_2, t) = B(l_u(x_1 + x_2) + t), \qquad \mathbf{x} = (x_1, x_2) \in \partial\Omega, t \in [0, T], \tag{25}$$

$$u(x_1, x_2, 0) = l_u(x_1 + x_2), \qquad \mathbf{x} = (x_1, x_2) \in \Omega, \tag{26}$$

where the thermal conductivity $K(\mathbf{x}) = (1 + x_1 + x_2)^2$, the thermal source loading $Q(\mathbf{x}, t) = (4l_u - 1)(1 + x_1 + x_2)$, the product of the density and the specific heat $\rho c = 1 + x_1 + x_2$. The analytical solution is $u(x_1, x_2, t) = l_u(x_1 + x_2) + t$. By using Laplace transformation and variable transformation $v^{(L)}(\mathbf{x}, p) = \sqrt{K(\mathbf{x})} u^{(L)}(\mathbf{x}, p)$, the transient heat conduction problem can be converted to

$$(\Delta - p) v^{(L)}(\mathbf{x}, p) = Q_g^{(L)}(\mathbf{x}, p), \quad \mathbf{x} \in \Omega, \tag{27}$$

$$Bv^{(L)}(x_1, x_2, p) = B(1 + x_1 + x_2)(l_u(x_1 + x_2) + 1/p^2), \quad \mathbf{x} = (x_1, x_2) \in \partial\Omega, \tag{28}$$

where the generalized heat source $Q_g^{(L)}(\mathbf{x}, p) = (4l_u - 1 - l_u(x_1 + x_2))(1 + x_1 + x_2)$. To vanish this generalized heat source term, the associated differential operator $\mathfrak{R}_1^{(L)} = \Delta - \dfrac{\Delta Q_g^{(L)}(\mathbf{x}, p)}{Q_g^{(L)}(\mathbf{x}, p)} = \Delta + \dfrac{4l_u}{(4l_u - 1 - l_u(x_1 + x_2))(1 + x_1 + x_2)}$ is implemented here. Then the related T-complete functions ϕ_k^i and φ_k^i [13] can be derived as follows:

$$\phi_k^i = \Big\{ I_0\big(\kappa_1 r_k^i\big), I_1\big(\kappa_1 r_k^i\big)\cos\big(\theta_k^i\big), I_1\big(\kappa_1 r_k^i\big)\sin\big(\theta_k^i\big), \dots,$$
$$I_{N_T/2}\big(\kappa_1 r_k^i\big)\cos\big(N_T\theta_k^i/2\big), I_{N_T/2}\big(\kappa_1 r_k^i\big)\sin\big(N_T\theta_k^i/2\big) \Big\} \tag{29a}$$

$$\varphi_k^i = \Big\{ J_0\big(\kappa_2 r_k^i\big), J_1\big(\kappa_2 r_k^i\big)\cos\big(\theta_k^i\big), J_1\big(\kappa_2 r_k^i\big)\sin\big(\theta_k^i\big), \dots,$$
$$J_{N_T/2}\big(\kappa_2 r_k^i\big)\cos\big(N_T\theta_k^i/2\big), J_{N_T/2}\big(\kappa_2 r_k^i\big)\sin\big(N_T\theta_k^i/2\big) \Big\} \tag{29b}$$

where $\kappa_1 = \sqrt{p}$ and $\kappa_2 = \sqrt{\dfrac{4l_u}{(4l_u - 1 - l_u(x_1 + x_2))(1 + x_1 + x_2)}}$, $r_k^i = \sqrt{(x_{k1}^i - \tilde{x}_1^i)^2 + (x_{k2}^i - \tilde{x}_2^i)^2}$,

$\theta_k^i = \arctan\left(\dfrac{x_{k2}^i - \tilde{x}_2^i}{x_{k1}^i - \tilde{x}_1^i}\right)$, in which $x_k^i = (x_{k1}^i, x_{k2}^i)$, $\tilde{x}^i = (\tilde{x}_1^i, \tilde{x}_2^i)$.

Under several parameter settings, Table 1 presents the Lerr errors obtained by the proposed LCTM+FTA for the transient heat conduction problems with full Dirichlet BCs and mixed BCs in comparison with the exact Laplace-space solutions + FTA. In the proposed LCTM, 1020 uniform-distributed nodes are used. In the mixed BC cases, only the boundary ($x_1 = 0, 0 \le x_2 \le 1$) is imposed on Neumann boundary condition, the remaining boundaries are imposed on Dirichlet boundary conditions. From Table 1, it can be found that with different parameter settings, the proposed LCTM + FTA provide the satisfactory numerical solutions for both the full Dirichlet BC cases and the mixed BC cases, which have a slight loss of accuracy in comparison with the retrieved time-dependent solutions from the exact Laplace-space solutions by fixed Talbot algorithm. It reveals that the numerical error is mainly

Table 1: Lerr errors obtained by the proposed LCTM + FTA for the transient heat conduction problems with full Dirichlet BCs and mixed BCs in comparison with the exact Laplace-space solutions + FTA under several parameter settings.

Parameter setting	Lerr errors		
	Full Dirichlet BCs	Mixed BCs	Exact Laplace-space solutions
$l_u = 0.1, T = 10$	1.39E − 2	1.56E − 2	1.07E − 2
$l_u = 0.1, T = 100$	1.41E − 3	1.57E − 3	1.08E − 3
$l_u = 0.1, T = 500$	2.82E − 4	3.15E − 4	2.17E − 4
$l_u = 0.01, T = 10$	4.88E − 3	6.96E − 3	1.08E − 3
$l_u = 0.01, T = 100$	4.91E − 4	7.00E − 4	1.08E − 4
$l_u = 0.01, T = 500$	9.84E − 5	1.40E − 4	2.16E − 5

appeared in the ill-posed NILT process. Moreover, the LCTM solutions for the full Dirichlet BC cases are slight accurate than the ones for the mixed BC cases. Fig. 1 displays the error distributions of the proposed LCTM+FTA for transient heat conduction problems with $l_u = 0.01$ subjected to full Dirichlet BCs and mixed BCs at different time instants (T = 10, 100, 500). It can be observed from Fig. 1 that the maximum relative errors are appeared at the central region of the computational domains for full Dirichlet BC cases and the center region of the left boundary of the computational domains for mixed BC cases. Similar to the conclusions drawn from Table 1, the LCTM solutions for the full Dirichlet BC cases are slight accurate than the ones for the mixed BC cases.

4 CONCLUSIONS

This paper proposes a novel localized collocation solver based on the localized collocation Trefftz method (LCTM) in conjunction with fixed Talbot algorithm (FTA) and the generalized reciprocity method (GRM) for transient heat conduction analysis in heterogeneous materials under thermal loading. The transient heat conduction analysis in heterogeneous materials with arbitrary spatial variations can be carried out by implementing the proposed LCTM coupled with the GRM, where the proposed scheme provides more accurate and efficient solutions than those obtained by the traditional FEM due to the use of the semi-analytical T-complete functions, and it is available for transient heat conduction analysis in heterogeneous materials without the related derived T-complete functions. The satisfactory numerical solutions can be obtained by using the proposed LCTM + FTA, which have a slight loss of accuracy in comparison with the retrieved time-dependent solutions from the exact Laplace-space solutions by fixed Talbot algorithm. It reveals that the numerical error is mainly appeared in the ill-posed NILT process. Of course, it would be interesting to develop an appropriate NILT to make the proposed localized collocation solver more efficient for transient heat conduction analysis in heterogeneous materials under thermal loading, and investigates the efficiency of the proposed scheme for solving the transient heat conduction problems without the close-form analytical solutions, which are now under intense study.

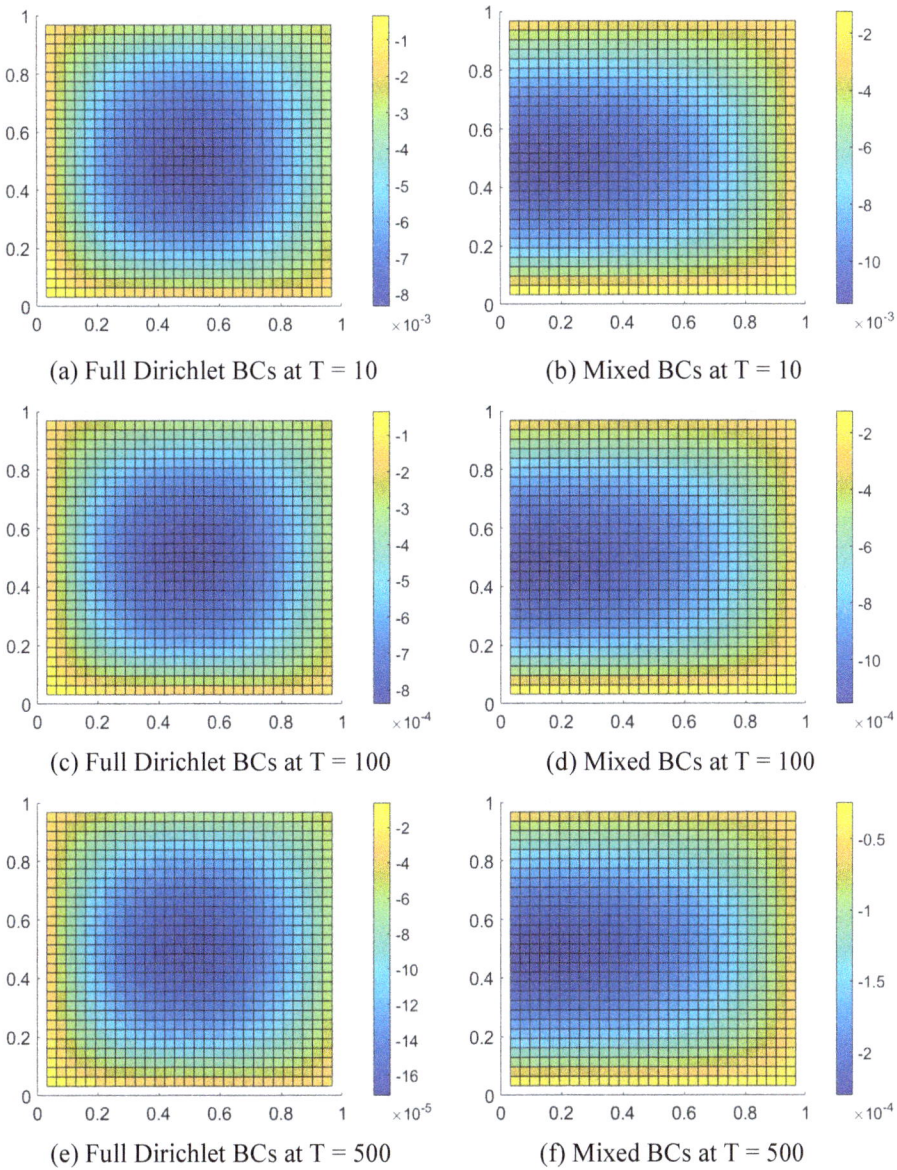

(a) Full Dirichlet BCs at T = 10   (b) Mixed BCs at T = 10

(c) Full Dirichlet BCs at T = 100   (d) Mixed BCs at T = 100

(e) Full Dirichlet BCs at T = 500   (f) Mixed BCs at T = 500

Figure 1:   Error distributions of the proposed LCTM + FTA for transient heat conduction problems with full Dirichlet BCs and mixed BCs at different time instants (T = 10, 100, 500).

ACKNOWLEDGEMENTS

The work described in this paper was supported by the National Science Funds of China (Grant No. 11772119), the Fundamental Research Funds for the Central Universities (Grant No. B200202124), the Foundation for Open Project of State Key Laboratory of Mechanics and Control of Mechanical Structures (Nanjing University Of Aeronautics And Astronautics)

(Grant No. MCMS-E-0519G01), and Six Talent Peaks Project in Jiangsu Province of China (Grant No. 2019-KTHY-009).

REFERENCES

[1]   Malek, M., Izem, N. & Seaid, M., A three-dimensional enriched finite element method for nonlinear transient heat transfer in functionally graded materials. *International Journal of Heat and Mass Transfer*, **155**, p. 119804, 2020.

[2]   Feng, W.Z. & Gao, X.W., An interface integral equation method for solving transient heat conduction in multi-medium materials with variable thermal properties. *International Journal of Heat and Mass Transfer*, **98**, pp. 227–239, 2016.

[3]   Sladek, J., Sladek, V., Krivacek, J. & Zhang, C., Local BIEM for transient heat conduction analysis in 3-D axisymmetric functionally graded solids. *Computational Mechanics*, **32**(3), pp. 169–176, 2003.

[4]   Xi, Q., Fu, Z., Wu, W., Wang, H. & Wang, Y., A novel localized collocation solver based on Trefftz basis for potential-based inverse electromyography. *Applied Mathematics and Computation*, **390**, p. 125604, 2021.

[5]   Liu, Y.C., Fan, C.M., Yeih, W., Ku, C.Y. & Chu, C.L., Numerical solutions of two-dimensional Laplace and biharmonic equations by the localized Trefftz method. *Computers & Mathematics with Applications*, **88**, pp. 120–134, 2021.

[6]   Sutradhar, A. & Paulino, G.H., The simple boundary element method for transient heat conduction in functionally graded materials. *Computer Methods in Applied Mechanics and Engineering*, **193**(42–44), pp. 4511–4539, 2004.

[7]   Abate, J. & Whitt, W., A unified framework for numerically inverting Laplace transforms. *INFORMS Journal on Computing*, **18**(4), pp. 408–421, 2006.

[8]   Fu, Z.J., Yang, L.W., Xi, Q. & Liu, C.S., A boundary collocation method for anomalous heat conduction analysis in functionally graded materials. *Computers & Mathematics with Applications*, **88**, pp. 91–109, 2021.

[9]   Fu, Z.J., Xi, Q., Chen, W. & Cheng, A.H.D., A boundary-type meshless solver for transient heat conduction analysis of slender functionally graded materials with exponential variations. *Computers & Mathematics with Applications*, **76**(4), pp. 760–773, 2018.

[10]   Partridge, P.W., Brebbia, C.A. & Wrobel, L.C., *The Dual Reciprocity Boundary Element Method*, Computational Mechanics Publications, 1992.

[11]   Nowak, A.J. & Neves, A.C., *The Multiple Reciprocity Boundary Element Method*, Computational Mechanics Publication, 1994.

[12]   Gavete, L., Gavete, M.L. & Benito, J.J., Improvements of generalized finite difference method and comparison with other meshless method. *Applied Mathematical Modelling*, **27**(10), pp. 831–847, 2003.

[13]   Li, Z.C., Lu, T.T., Huang, H.T. & Cheng, A.H.D., Trefftz, collocation, and other boundary methods – A comparison. *Numerical Methods for Partial Differential Equations: An International Journal*, **23**(1), pp. 93–144, 2007.

SECTION 3
ENGINEERING AND INDUSTRIAL APPLICATIONS

BEM-BASED FORMULATION FOR THE ANALYSIS OF MULTI-STORY BUILDINGS INCLUDING SOIL-STRUCTURE INTERACTION

AHMED U. ABDELHADY[1,2], AHMED FADY FARID[2], YOUSSEF F. RASHED[2] & JASON MCCORMICK[1]
[1]Department of Civil and Environmental Engineering, University of Michigan, USA
[2]Department of Structural Engineering, Cairo University, Egypt

ABSTRACT

Modeling and analyzing multi-story buildings is an important part of structural engineering. Typically, this analysis is carried out using the finite element method by assembling the stiffness matrices of the floor elements and the vertical supporting elements at the intersecting nodes. Any consideration of soil-structure interaction (SSI) is often simplified using a Winkler model. However, the procedure for modeling practical buildings of complex geometries using the finite element method can be cumbersome. Alternatively, a new formulation that is based on the boundary element method (BEM) is presented that provides a seamless procedure for modeling practical buildings as the discretization of the floors is done at the floor perimeter and SSI is modeled using an elastic half space (EHS) model. The stiffness matrices of the slab and raft are generated using the BEM by introducing an additional collocation scheme at their intersection with the columns and the underlying soil. Columns are modeled as skeletal frame elements and the floors are considered as rigid diaphragms in their planes. Soil is modeled as an EHS and its stiffness matrix is derived based on the Bousinessq solution of an elastic, isotropic, homogenous, and infinite thickness half space. Assembly of the overall building stiffness matrix is carried out using the well-known assembly procedure associated with the stiffness analysis method. The proposed methodology is validated by comparing the results against the more traditional finite element approach. An illustrative example is solved showing agreement of the results between the proposed methodology and the finite element method.

Keywords: boundary element method, elastic half space, soil–structure interaction, multi-story buildings, stiffness analysis.

1 INTRODUCTION

There is a need to develop robust techniques for modeling and analyzing multi-story buildings. This need provides an opportunity for the use of the boundary element method, as a meshless technique, in modeling multi-story buildings while accounting for soil-structure interaction. Multi-story buildings consist of horizontal elements (e.g., slabs, beams, etc.) and vertical elements (e.g., columns, walls, etc.). The horizontal and vertical elements are supported above the ground by the foundation (e.g., raft, isolated footings, etc.).

The boundary element formulation of a flat plate supported by columns only is presented in [1]. Many researchers also worked on modeling a flat plate supported by beams using the boundary element method either based on Kirchhoff–Love plate theory [2], [3] or Mindlin–Reissner plate theory [4], [5]. Edge beams are considered in the formulation presented by [2], [6] while in [5], [7] beams are modeled as a plate region with different thickness and material properties. In [4], a practical boundary element formulation is presented that can account for beams with any arbitrary configuration.

The problem of soil-structure interaction has been investigated extensively over the past few decades [8]–[12]. The analysis of a Mindlin–Reissner plate on an elastic half space (EHS) is presented in [8]. This analysis has been extended in [9] to account for the non-linearity of the soil using an iterative procedure. A similar approach is presented in [10] to

WIT Transactions on Engineering Sciences, Vol 131, © 2021 WIT Press
www.witpress.com, ISSN 1743-3533 (on-line)
doi:10.2495/BE440071

solve for a tensionless foundation. The analysis of the piled raft is presented in [11] where the pile-soil-raft interaction is considered.

In this work, a comprehensive methodology for the analysis of multi-story buildings including soil-structure interaction is presented. The presented methodology combines the use of the boundary element method to model the slabs and foundation raft, the elastic half space to model the underlying soil, and the stiffness analysis method. The presented methodology is applied to a three-story building and the results are validated using the finite element method.

2 STIFFNESS ANALYSIS OF MULTI-STORY BUILDINGS

Stiffness analysis has been widely used to analyze multi-story buildings under various loading conditions. The analysis applies Hooke's law which is stated as follows:

$$\{\mathbf{P}\} = [\mathbf{K}]\{\mathbf{u}\}, \tag{1}$$

where $\{\mathbf{P}\}$ is the loading vector; $\{\mathbf{u}\}$ is the displacement vector; and $[\mathbf{K}]$ is the total stiffness matrix.

The total stiffness matrix $[\mathbf{K}]$ for the structure shown in Fig. 1 is obtained by assembling the stiffness matrices of the slabs, columns, raft, and soil. Columns are modeled using frame elements and their stiffness matrices are obtained as described in [13]. The slabs and raft are modeled using the Mindlin–Reissner plate bending theory while the supporting soil below the building raft is modeled as an elastic half space.

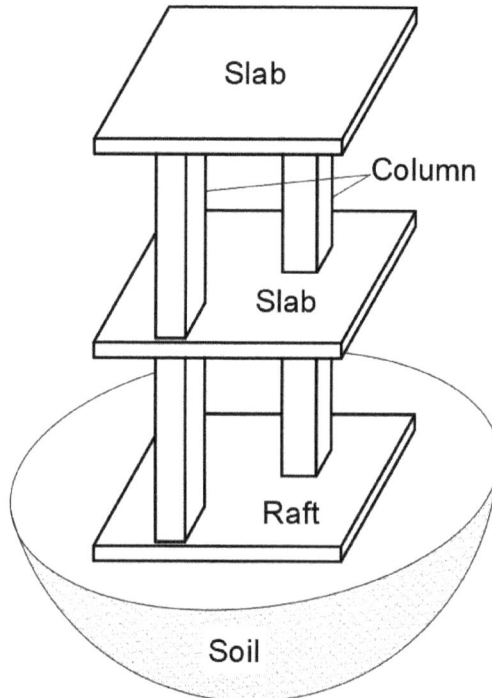

Figure 1: Schematic of the multi-story building under consideration.

3 SOIL STIFFNESS MATRIX

The soil below the raft is modeled as an elastic half space (EHS), Fig. 2, and the stiffness matrix of the elastic half space is estimated as follows:

$$[\mathbf{K}_{EHS}] = [\mathbf{F}_{EHS}]^{-1}, \tag{2}$$

where $[\mathbf{F}_{EHS}]$ is the flexibility matrix of the elastic half space.

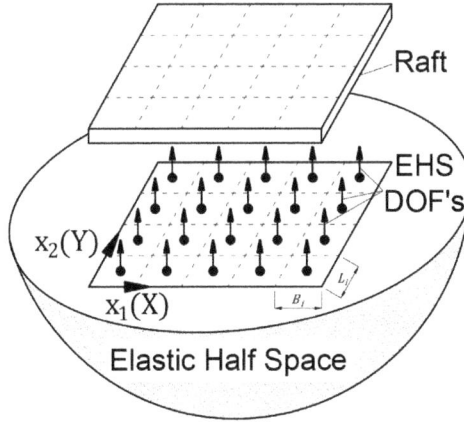

Figure 2: Raft on elastic half space.

The coefficients of $[\mathbf{F}_{EHS}]$ are calculated based on the Bousinessq solution [14] which provides the displacement of a point (j) lying on the surface of an elastic, isotropic, homogenous, and infinite thickness half space due to a concentrated unit load at point (i),

$$f_{ij} = \frac{1-v}{2\pi G r_{ij}} \qquad \forall \, i \neq j, \tag{3}$$

where f_{ij} represents the coefficients of $[\mathbf{F}_{EHS}]$; v is the Poisson's ratio; $G = E/2(1 + v)$ is the modulus of rigidity; E is the modulus of elasticity; and r_{ij} is the distance between the loading point (i) and the displacement point (j).

It should be noted that calculating the diagonal terms of $[\mathbf{F}_{EHS}]$ using eqn (3) (i.e., calculating the displacement at the concentrated load point) will lead to a singularity. This singularity is handled by replacing the concentrated load with an equivalent pressure that is distributed over the ith cell area ($B_i \times L_i$) [8].

$$f_{ii} = \frac{R_i(1-v)}{\pi G B_i}, \tag{4}$$

where R_i is the rectangularity factor of the ith cell which is calculated as follows:

$$R_i = \ln\left[\beta_i^{-\beta_i}\left(1 + \sqrt{1 + \beta_i^2}\right)^{\beta_i}\left(\beta_i + \sqrt{1 + \beta_i^2}\right)\right], \tag{5}$$

where $\beta_i = B_i/L_i$ is the rectangularity ratio.

4 MINDLIN–REISSNER PLATE STIFFNESS MATRIX

The building slabs and raft are modeled using the Mindlin–Reissner plate bending theory. The boundary element method is used to solve the Mindlin–Reissner plate problem in order to generate the slabs' and raft's stiffness matrices.

4.1 Boundary integral equation for a plate with internal supports

Figs 3 and 4 show a boundary element model for a slab and raft, respectively. The indicial notation is used in this section where the Greek indices vary from 1 to 2 (to denote the x and y directions) and Roman indices vary from 1 to 3 (to denote the x, y, and the z directions). Slabs are modeled using Mindlin–Reissner plate bending theory where Γ and Ω are the slab boundary and domain, respectively. The intersection area between the plate and the internal supports (i.e., columns and soil) are called supporting cells. Each supporting cell has a node located in its centroid. The direct boundary integral equation for a slab with supporting cells can be written as follows [4]:

$$C_{ij}(\boldsymbol{\xi})u_j(\boldsymbol{\xi}) + \int_{\Gamma(\mathbf{x})} T_{ij}(\boldsymbol{\xi},\mathbf{x})u_j(\mathbf{x})\,d\Gamma(\mathbf{x}) - \int_{\Gamma(\mathbf{x})} U_{ij}(\boldsymbol{\xi},\mathbf{x})t_j(\mathbf{x})\,d\Gamma(\mathbf{x}) =$$
$$\sum_{N_s} \left[\int_{\Omega(s)} \left[U_{ik}(\boldsymbol{\xi},\mathbf{s}) - \frac{v}{(1-v)\lambda^2} U_{i\alpha,\alpha}(\boldsymbol{\xi},\mathbf{s})\delta_{3k} \right] F_k(s)\,d\Omega(s) \right], \quad (6)$$

where C_{ij} is the jump term; $\boldsymbol{\xi}$ is the source boundary point; $\mathbf{x}$ is the field point; T_{ij} and U_{ij} are the fundamental solution kernels for traction and displacement, respectively; t_j and u_j are the boundary generalized traction and displacement, respectively; N_s is the number of supporting cells; $\mathbf{s}$ is the centroid of the supporting cell; $\Omega(s)$ is the domain of the supporting cell; v is the Poisson's ratio; λ is the shear factor; F_k is the interaction force between the supporting cell and the plate per unit area.

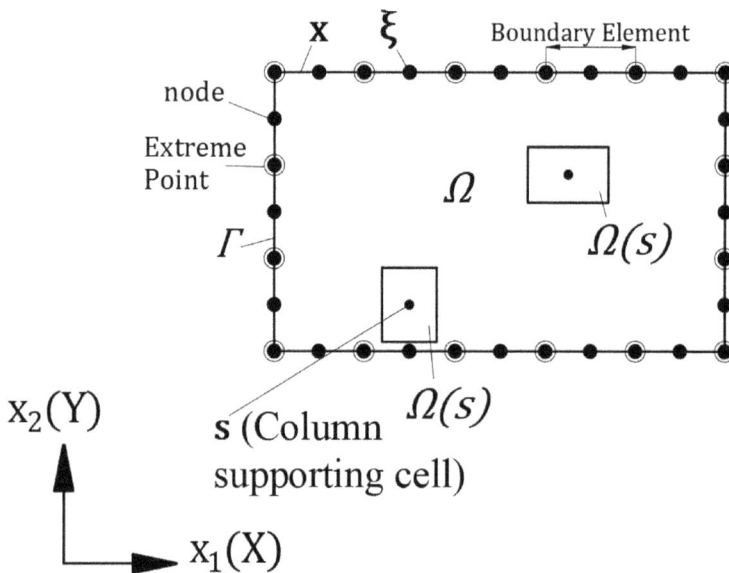

Figure 3: Slab boundary element discretization.

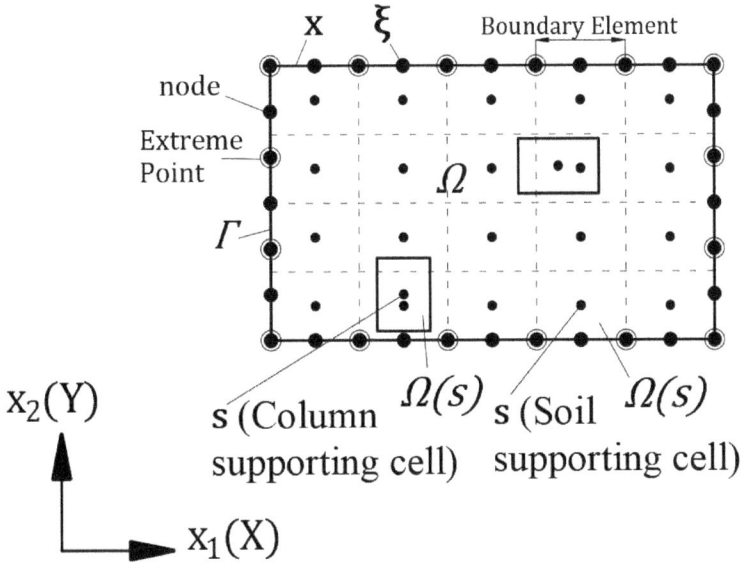

Figure 4: Raft boundary element discretization.

The plate boundary, Figs 3 and 4, is discretized into N_e quadratic boundary elements. Then, the traditional collocation procedure is carried out at each boundary node which leads to the following matrix form of eqn (6):

$$[\mathbf{A}(\boldsymbol{\xi},\mathbf{x})]_{3N\times 3N}\{\mathbf{u}/\mathbf{t}(\mathbf{x})\}_{3N\times 1} = [\mathbf{B}(\boldsymbol{\xi},\mathbf{s})]_{3N\times 3N_s}\{\mathbf{F}(\mathbf{s})\}_{3N_s\times 1}, \qquad (7)$$

where $[\mathbf{A}(\boldsymbol{\xi},\mathbf{x})]$ and $[\mathbf{B}(\boldsymbol{\xi},\mathbf{s})]$ are the coefficient matrices obtained from the left hand side and right hand side of eqn (6), respectively; $\{\mathbf{u}/\mathbf{t}(\mathbf{x})\}$ is the vector of the unknown boundary displacements or tractions; $\{\mathbf{F}(\mathbf{s})\}$ is the vector of the forces at the supporting cells; N is the number of the boundary nodes which equals $2N_e$.

The unknowns in eqn (7) are $\{\mathbf{u}/\mathbf{t}(\mathbf{x})\}$ and $\{\mathbf{F}(\mathbf{s})\}$. The number of unknowns $(3N + 3N_s)$ in eqn (7) is greater than the number of equations $(3N)$. A solution strategy that is based on an additional collocation scheme pattern is proposed to solve eqn (7).

4.2 Solution strategy

The solution strategy uses an additional collocation scheme between the centroid of the supporting cells $(\mathbf{s})$ as the source point and the slab boundary $(\mathbf{x})$ as the field point.

$$u_j(\mathbf{s}) + \int_{\Gamma(\mathbf{x})} T_{ij}(\mathbf{s},\mathbf{x})u_j(\mathbf{x})d\Gamma(\mathbf{x}) - \int_{\Gamma(\mathbf{x})} U_{ij}(\mathbf{s},\mathbf{x})t_j(\mathbf{x})d\Gamma(\mathbf{x}) = \sum_{N_s}\left[\int_{\Omega(s)}\left[U_{ik}(\mathbf{s},\mathbf{s}) - \frac{\nu}{(1-\nu)\lambda^2}U_{i\alpha,\alpha}(\mathbf{s},\mathbf{s})\delta_{3k}\right]F_k(\mathbf{s})d\Omega(\mathbf{s})\right]. \quad (8)$$

Eqn (8) is then written in the following matrix form:

$$\{\mathbf{u}(\mathbf{s})\}_{3N_s\times 1} + [\mathbf{A}(\mathbf{s},\mathbf{x})]_{3N_s\times 3N_s}\{\mathbf{u}/\mathbf{t}(\mathbf{x})\}_{3N_s\times 1} = [\mathbf{B}(\mathbf{s},\mathbf{s})]_{3N_s\times 3N_s}\{\mathbf{F}(\mathbf{s})\}_{3N_s\times 1}, \quad (9)$$

where $[\mathbf{A}(\mathbf{s}, \mathbf{x})]$ and $[\mathbf{B}(\mathbf{s}, \mathbf{s})]$ are the coefficient matrices obtained from the left hand side and right hand side of eqn (8), respectively.

The additional collocation scheme generates additional equations that are combined with eqn (7) to balance the number of equations with the number of unknowns as follows:

$$\begin{bmatrix} [\mathbf{A}(\boldsymbol{\xi}, \mathbf{x})]_{3N \times 3N} & -[\mathbf{B}(\boldsymbol{\xi}, \mathbf{s})]_{3N \times 3N_s} \\ [\mathbf{A}(\mathbf{s}, \mathbf{x})]_{3N_s \times 3N_s} & -[\mathbf{B}(\mathbf{s}, \mathbf{s})]_{3N_s \times 3N_s} \end{bmatrix} \begin{Bmatrix} \{\mathbf{u}/\mathbf{t}(\mathbf{x})\}_{3N \times 1} \\ \{\mathbf{F}(\mathbf{s})\}_{3N_s \times 1} \end{Bmatrix} = \begin{Bmatrix} \mathbf{0}_{3N \times 1} \\ -\{\mathbf{u}(\mathbf{s})\}_{3N_s \times 1} \end{Bmatrix}. \quad (10)$$

4.3 Stiffness matrix generation

The slab and raft stiffness matrices are obtained by setting each generalized displacement in $\{\mathbf{u}(\mathbf{s})\}$ that corresponds to the degrees of freedom at the supporting cells to unity (one at a time). Eqn 10 is then solved. The resulting $\{\mathbf{F}(\mathbf{s})\}$ represents a column (or row) in the slab or raft stiffness matrix. The following equation summarizes this process:

$$\begin{bmatrix} [\mathbf{A}(\boldsymbol{\xi}, \mathbf{x})]_{3N \times 3N} & -[\mathbf{B}(\boldsymbol{\xi}, \mathbf{s})]_{3N \times 3N_s} \\ [\mathbf{A}(\mathbf{s}, \mathbf{x})]_{3N_s \times 3N_s} & -[\mathbf{B}(\mathbf{s}, \mathbf{s})]_{3N_s \times 3N_s} \end{bmatrix} \begin{Bmatrix} [\mathbf{u}/\mathbf{t}(\mathbf{x})]_{3N \times 3N_s} \\ [\mathbf{K}_{slab}]_{3N_s \times 3N_s} \end{Bmatrix} = \begin{Bmatrix} \mathbf{0}_{3N \times 3N_s} \\ \mathbf{I}_{3N_s \times 3N_s} \end{Bmatrix}, \quad (11)$$

where $\mathbf{I}$ is the identity matrix; and $\mathbf{K}_{slab}$ is the slab stiffness matrix.

5 APPLICATION

In this section, the presented methodology is applied to analyze the multi-story building shown in Fig. 5. Analysis results are then compared with the finite element method (FEM) to validate the presented boundary element method approach.

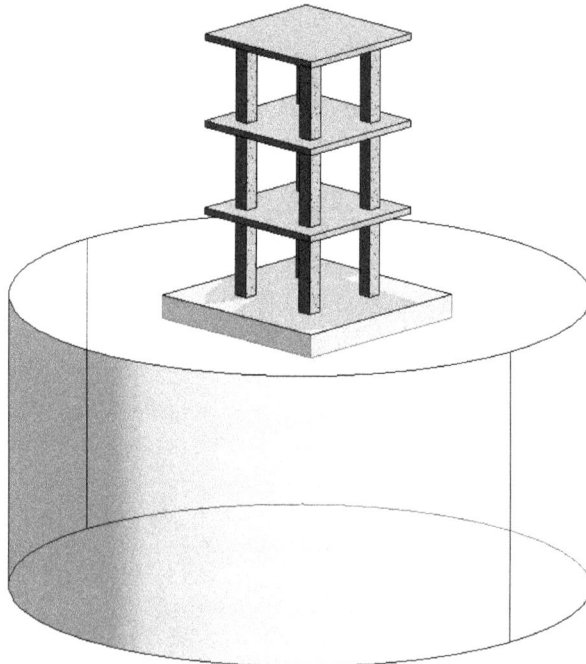

Figure 5: Structural system of the multi-story building.

5.1 Model description

The considered multi-story building consists of three levels as shown in Fig. 5. Dimensions of the slab are 5×5 m^2 and its thickness is 0.25 m. The slabs are supported by four columns which have a cross-section of 0.5×0.5 m^2. The four columns extend from the top slab to the raft which is 7×7 m^2 and its thickness is 0.8 m.

The material used for the slabs, columns, and raft is concrete ($E = 2.5 \times 10^6$ t/m^2 and $\nu = 0.2$). The modulus elasticity of the soil is 10^3 t/m^2 while the Poisson's ratio is taken as 0.2. The three slabs are loaded with a uniform load of 1.6 t/m^2.

The boundary element models of a slab and raft are shown in Fig. 6. Fig. 7 shows the finite element model used to validate the results from the presented methodology [15]. The slabs and raft are modeled using Reissener plates, columns are modeled using frame elements, and the soil is modeled using solid elements. The size of the modeled soil block is $27 \times 27 \times 14$ m^3.

5.2 Results and discussion

Figs 8 and 9 compare the deflection (u_3) obtained from the presented boundary element method approach and that from the finite element model for the top slab and the raft, respectively. The maximum deflection of the raft in the FEM is 19 mm which is around 5% less than the maximum deflection in the BEM (20 mm). For the top slab, the difference between the maximum and minimum deflection is 0.2 mm in the BEM which matches the results from the FEM with less than 1% difference.

Figs 10 and 11 compare the bending moment (M_{11}) obtained from the BEM and the FEM for the top slab and the raft, respectively. The maximum negative moment, M_{11}, for the middle of the top slab in the BEM is -0.731 t. m which is approximately 1.5% larger than the FEM (-0.72 t. m). The maximum positive, M_{11}, for the top slab (at the column edge) in the BEM is 1.2 t. m which is 4% less than the FEM (1.25 t. m). For the raft, Fig. 11, maximum negative and positive M_{11} in the BEM and FEM vary within 2–5% as well. These results validate the presented methodology as they show agreement between the BEM and the FEM.

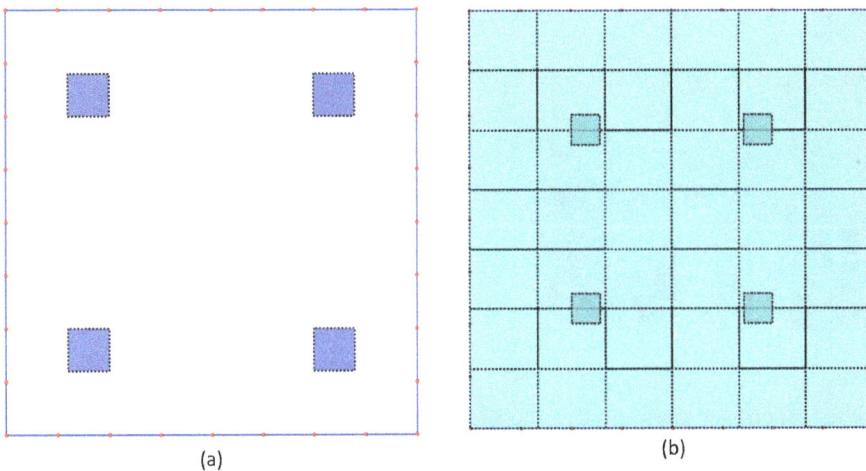

(a)   (b)

Figure 6: Boundary element model. (a) Slab; and (b) Raft.

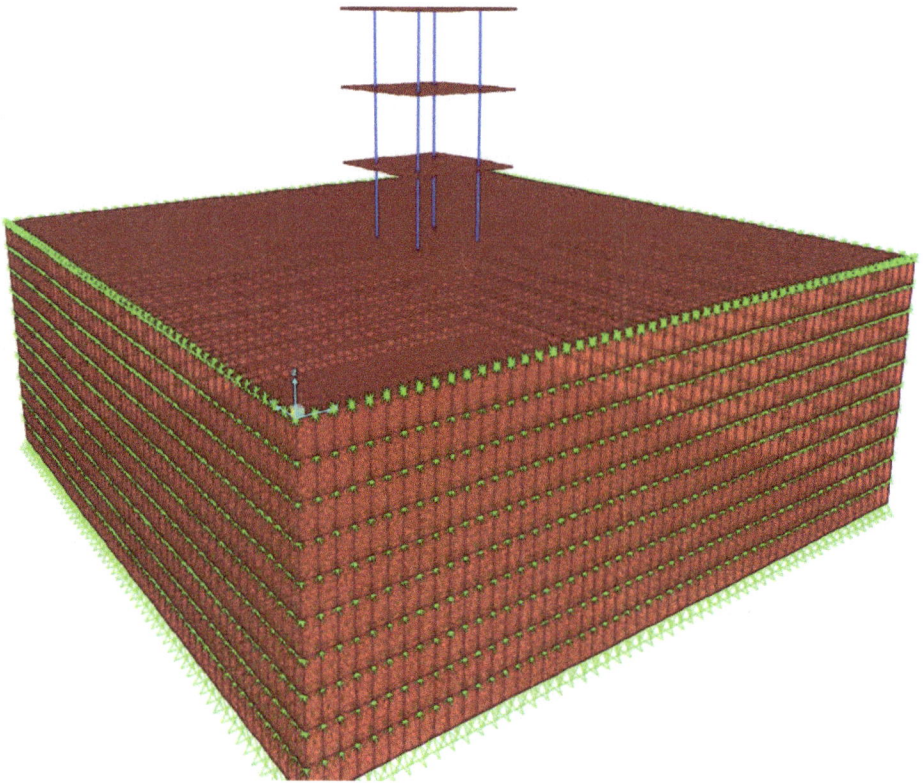

Figure 7: Three-dimensional finite element model for the building and the supporting soil.

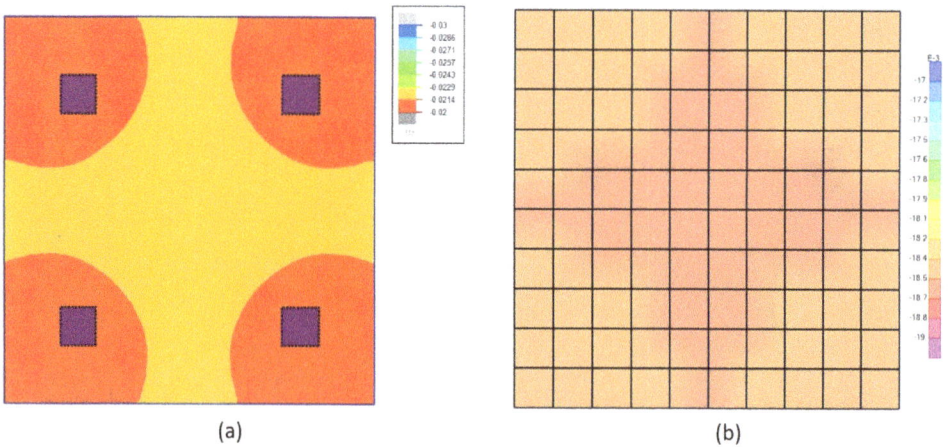

Figure 8:   Top slab deflection (u_3). (a) Presented methodology; and (b) Finite element method.

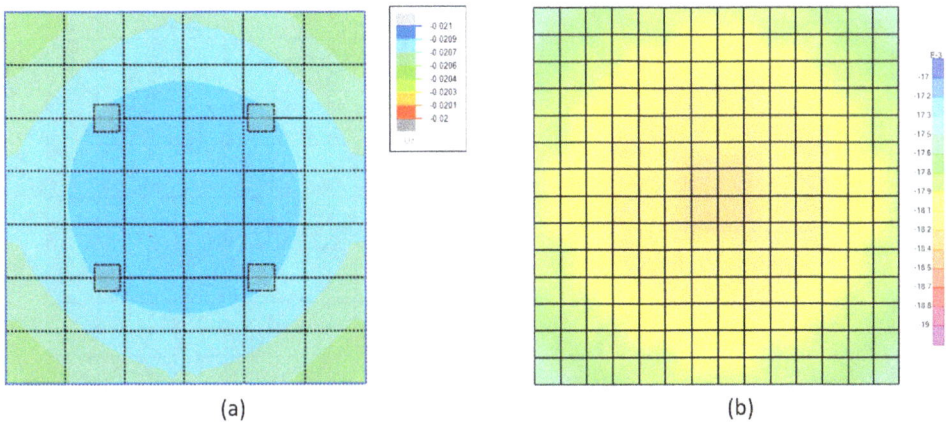

Figure 9: Raft deflection (u_3). (a) Presented methodology; and (b) Finite element method.

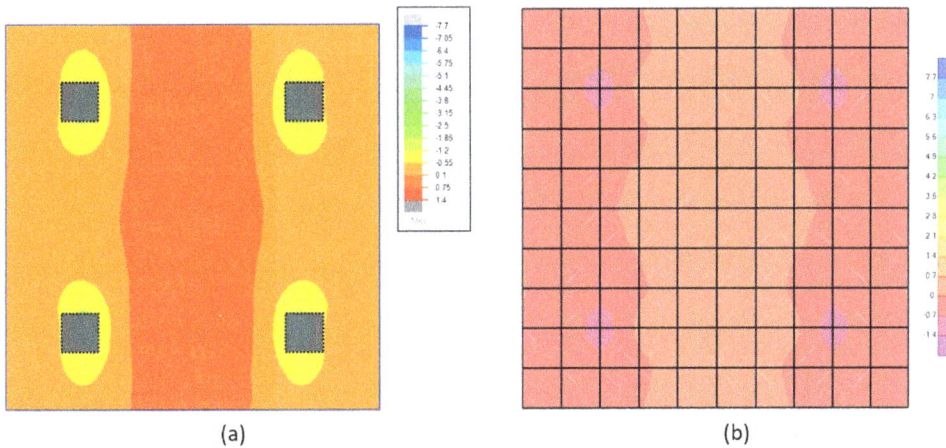

Figure 10: Top slab bending moment (M_{11}). (a) Presented methodology; and (b) Finite element method.

6 CONCLUSIONS

This paper presents a methodology for the analysis of multi-story buildings including soil-structure interaction. The slabs and raft are modeled using a Mindlin–Reissener plate and their stiffness matrices are obtained using the boundary element method. Columns are modeled as frame elements and the soil is modeled as an elastic half space. The interaction between the soil and the building raft is handled using an internal support BEM-based formulation for the Mindlin–Reissener plate. Finally, the stiffness matrices of all building elements as well as the soil are assembled using the well-known technique of stiffness analysis method. The presented methodology is validated by comparing its results against the finite element method.

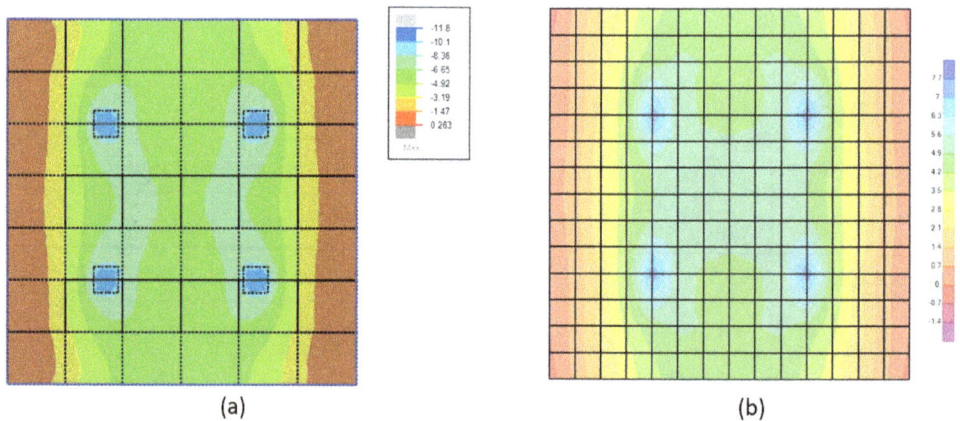

Figure 11: Raft bending moment (M_{11}). (a) Presented methodology; and (b) Finite element method.

REFERENCES

[1]   Rashed, Y.F., Boundary element modelling of flat plate floors under vertical loading. *International Journal for Numerical Methods in Engineering*, **62**(12), pp. 1606–1635, 2005.

[2]   Hu, C. & Hartley, G.A., Elastic analysis of thin plates with beam supports. *Engineering Analysis with Boundary Elements*, **13**, pp. 229–238, 1994.

[3]   de Paiva, J.B., Boundary element formulation of building slabs. *Engineering Analysis with Boundary Elements*, **17**, pp. 105–110, 1996.

[4]   Abdelhady, A.U. & Rashed, Y.F., A practical boundary element analysis of slab-beam floor type. *Engineering Analysis with Boundary Elements*, **97**, pp. 23–38, 2018.

[5]   Fernandes, G.R. & Konda, D.H., A BEM formulation based on Reissner's hypothesis for analysing the coupled stretching-bending problem of building floor structures. *Engineering Analysis with Boundary Elements*, **36**, pp. 1377–1388, 2012.

[6]   Sapountzakis, J.T. & Katsekadelis, J.T., Analysis of plates reinforced with beams. *Computational Mechanics*, **26**, pp. 66–74, 2000.

[7]   Fernandes, G.R., A BEM formulation for linear bending analysis of plates reinforced by beams considering different materials. *Engineering Analysis with Boundary Elements*, **33**, pp. 1132–1140, 2009.

[8]   Shaaban, A.M. & Rashed, Y.F., A coupled BEM-stiffness matrix approach for analysis of shear deformable plates on elastic half space. *Engineering Analysis with Boundary Elements*, **37**, pp. 699–707, 2013.

[9]   Farid, A.F., Reda, M. & Rashed, Y.F., Efficient analysis of plates on nonlinear foundations. *Engineering Analysis with Boundary Elements*, **83**, pp. 1–24, 2017.

[10] Farid, A.F. & Rashed, Y.F., BEM for thick plates on unilateral Winkler springs. *Innovative Infrastructure Solutions*, **3**, p. 26, 2018.

[11] Shehata, O.E., Farid, A.F. & Rashed, Y.F., Practical boundary element method for piled rafts. *Engineering Analysis with Boundary Elements*, **97**, pp. 67–81, 2018.

[12] Azzam, O.A.A., Farid, A.F., Rashed, Y.F. & Elghazaly, H.A., The use of out-of-core iterative solvers for large 3D soil problems. *Engineering Analysis with Boundary Elements*, **118**, pp. 124–142, 2020.

[13]   Taranath, B.S., *Structural Analysis and Design of Tall Buildings*, McGraw-Hill: New York, USA, 1988.
[14]   Selvadurai, A.P.S., *Elastic Analysis of Soil Foundation Interaction*, Elsevier: Amsterdam, 1979.
[15]   CSI, *CSI Analysis Reference Manual*, Berkeley, California, 2016.

WIT Transactions on Engineering Sciences, Vol 131, © 2021 WIT Press
www.witpress.com, ISSN 1743-3533 (on-line)

PLATE BUCKLING INCLUDING EFFECTS OF SHEAR DEFORMATION AND PLATE BENDING CURVATURES USING THE BOUNDARY ELEMENT METHOD

LEANDRO PALERMO JR.[1], VITOR CAUÊ GOMES[1] & LUIZ CARLOS WROBEL[2]
[1]School of Civil Engineering, Architecture and Urban Design, State University of Campinas, Brazil
[2]Department of Civil Engineering, Pontifical Catholic University of Rio de Janeiro, Brazil

ABSTRACT

In this paper, the plate bending curvature was included in the geometrical non-linearity (GNL) effect beyond the deflection derivatives to perform plate buckling analyses. The boundary element method (BEM) was adopted and the formulation employed two integrals related to the GNL effect, with one computed on the boundary and the other on the domain. The eigenvalue problem was solved with the inverse iteration method. Results obtained with different boundary conditions were compared to values in the literature.

Keywords: plate buckling, Mindlin plate, Reissner plate, bucking parameter, tangential differential operator.

1 INTRODUCTION

In-plane forces affect the plate bending behavior when the deflection surface is considered in the plate equilibrium. The problem is geometrically nonlinear in the case of large deflections, when the stretching and the bending of the plate become coupled [1]. Buckling analysis is one of the ways to evaluate the effect of in-plane forces when deflections remain small and the in-plane forces can be related to in-plane tractions. Timoshenko and Woinowsky-Krieger [1] presented the equilibrium equation for the classical bending model with the geometric non-linearity (GNL) effect containing the derivatives of the plate deflection weighted by in-plane forces, which is used in buckling analyses. The inclusion of the effect of the shear deformation in the bending model improves the accuracy of the plate stresses computation [2] or the dynamical behavior of the plate [3]. The buckling load values present significant changes according to plate thickness values beyond the flexural rigidity when the effect of shear deformation is included in the bending model [4], with reference to results obtained with the classical model even considering the same GNL effect presented by Timoshenko in both models. Questions arise on the effect of shear deformation in the GNL when buckling analyses for moderately thick plates are performed. Dawe and Roufaeil [5] discussed the plate buckling analyses considering the effect of shear deformation following the first study presented by Herrmann and Armenakas [6]. The inclusion of the curvatures (the first derivative of rotations) beyond the deflection derivatives in the GNL effect was the main point in the discussion. Sun [7] presented a detailed and comprehensive study using the equations of motion for the Timoshenko beam with the curvature included in the GNL effect beyond the deflection derivative. Sun considered the Trefftz and the Biot theory in the analyses and showed the buckling loads were reduced in the range of intermediate wavelengths, i.e., when the shear deformation is pronounced. Brunelle and Robertson [8] presented two ways to obtain the differential equations for Mindlin plates under a general state of non-uniform initial stress considering the curvatures and the deflection derivatives in the GNL effect. Mizusawa [9] showed the effect of curvatures was greater for certain types of boundary conditions, whereas it was not significant for others when the derivatives of deflection would be enough for the analyses. Smith [10] proposed a finite element

WIT Transactions on Engineering Sciences, Vol 131, © 2021 WIT Press
www.witpress.com, ISSN 1743-3533 (on-line)
doi:10.2495/BE440081

formulation including both derivatives on displacements (deflection and rotations) in the buckling of thick plates. Doong [11] and Matsunaga [12] studied improvements to the model representing the effect of shear deformation in plate bending, where the buckling analyses considered the derivatives of rotations and deflections combined with a high-order theory representing the effect of shear deformation.

The boundary element formulation for buckling analyses derived in this study employs two integrals containing the GNL effect, with one computed on the domain and the other computed on the boundary. The first derivatives of the deflection and rotations were used in the kernels of integrals related to the GNL effect, and no relation was required for the derivatives of in-plane forces. The natural conditions of the buckling problem were considered and related to boundary integrals containing the GNL effect in case of free edges or when the deflection and/or rotations were released for certain types of boundary conditions, i.e., the hard or the soft condition. The present formulation contains the general GNL effect with reference to that presented in [4], where only the first derivatives of the deflection were considered. The numerical implementation employed quadratic shape functions to approximate displacements (deflections and rotations), distributed shears, and moments in the boundary elements whereas constant elements were used to discretize both integrals related to the GNL effect. Constant elements were the lower type of element to evaluate the behavior of the formulation containing the GNL effect. An algebraic manipulation using both integrals with the GNL effect was carried out to perform integrations only on the sides of cells inside the domain in problems with known displacements (deflection and rotations) on the plate boundary. The inverse iteration and Rayleigh quotient were used to compute the lowest eigenvalues with the corresponding eigenvectors. The changes in the value of the buckling parameter according to the plate thickness were compared to values available in the literature.

2 BOUNDARY INTEGRAL EQUATIONS

The constitutive equations are written next with a unified notation for the Reissner and Mindlin bending models. The Latin indices take on values {1, 2 and 3} and Greek indices take on values {1, 2}.

$$M_{\alpha\beta} = D \frac{(1-\nu)}{2} \left(u_{\alpha,\beta} + u_{\beta,\alpha} + \frac{2\nu}{1-\nu} u_{\gamma,\gamma} \delta_{\alpha\beta} \right) + \delta_{\alpha\beta} q RE, \qquad (1)$$

$$Q_\alpha = D \frac{(1-\nu)}{2} \lambda^2 (u_\alpha + u_{3,\alpha}), \qquad (2)$$

with

$$D = \frac{Eh^3}{12(1-\nu^2)} \quad \lambda^2 = 12 \frac{\kappa^2}{h^2}; \quad RE = \frac{\nu}{\lambda^2(1-\nu)},$$

where u_α is the plate rotation in direction α, and u_3 is the plate deflection, D is the flexural rigidity, h is the plate thickness, ν is Poisson's ratio, q is the distributed load on the plate domain and $\delta_{\alpha\beta}$ is the Kronecker delta. The product qRE in eqn (1) corresponds to the linearly weighted average effect of the normal stress component in the thickness direction, which should be considered in the Reissner model [2] but not in the Mindlin model [3] (RE = 0). This term is null in buckling problems because the distributed load q is equal to zero. The shear parameter is equal to 5/6 and $\pi^2/12$ for the Reissner and the Mindlin model, respectively, and it is the only difference introduced according to the model employed in the analysis.

The natural conditions and the equilibrium equations for the problem can be obtained with the calculus of variations [13], [14]. The energy functional of the plate is given by:

$$\Pi = \int_\Omega \left\{ \frac{D(1-v)}{4} \left[u_{\alpha,\beta}{}^2 + u_{\alpha,\beta} u_{\beta,\alpha} + \frac{2v}{(1-v)} u_{\gamma,\gamma}{}^2 + \lambda^2 \left(u_\alpha + u_{3,\alpha} \right)^2 \right] \right\} d\Omega + \cdots$$

$$+ \int_\Omega \frac{h^2}{24} \left(N_{\alpha\beta} u_{\gamma,\alpha} u_{\gamma,\beta} \right) d\Omega + \int_\Omega \frac{1}{2} \left(N_{\alpha\beta} u_{3,\alpha} u_{3,\beta} \right) d\Omega - \int_{\Gamma_f} (Pw + EM_\alpha u_\alpha) d\Gamma. \tag{3}$$

The energy functional of the plate was written in the complete form without the distributed load in eqn (3). The first integral (domain integral) is the strain energy whereas the GNL effect appeared in subsequent integrals containing the in-plane forces ($N_{\alpha\beta}$). The last integral is the potential energy of the external loads. EM_1 and EM_2 are couples in directions 1, 2 and P is the out-of-plane load distributed on a portion of the boundary (Γ_f). The displacements (u_1, u_2 and u_3) are not prescribed on the portion of the boundary line Γ_f. The energy functional of the plate can be written as a general function to be minimized with the calculus of variations:

$$\Pi = \int_\Omega F\left(u_1, u_2, u_3, u_{1,1}, u_{2,1}, u_{3,1}, u_{1,2}, u_{2,2}, u_{3,2} \right) d\Omega. \tag{4}$$

The Euler equations obtained from the minimization of eqn (4) are given by:

$$\frac{\partial F}{\partial u_i} - \frac{\partial}{\partial x_\alpha} \left(\frac{\partial F}{\partial u_{i,\alpha}} \right) = 0 \;{}_{(i=1,2,3)}.$$

The equilibrium equations are obtained when the constitutive equations are introduced in the resultant expressions from the application of Euler equations:

$$M_{\gamma\beta,\beta} - Q_\gamma + \frac{h^2}{12} \frac{\partial}{\partial x_\alpha} \left(N_{\alpha\beta} \frac{\partial u_\gamma}{\partial x_\beta} \right) = 0, \tag{5}$$

$$Q_{\alpha,\alpha} + \frac{\partial}{\partial x_\alpha} \left(N_{\alpha\beta} \frac{\partial u_3}{\partial x_\beta} \right) = 0. \tag{6}$$

The natural conditions introduce the requirements on the boundary portion (Γ_f) with not prescribed displacements where the variations on displacements are not null ($\delta u_i \neq 0$):

$$\left(\frac{\partial F}{\partial u_{i,\alpha}} n_\alpha \right) \delta u_i = 0 \xrightarrow{yields} \frac{\partial F}{\partial u_{i,\alpha}} n_\alpha = 0,$$

$$t_\gamma = EM_\gamma - \frac{h^2}{12} \left(n_\alpha N_{\alpha\beta} u_{\gamma,\beta} \right), \tag{7}$$

$$t_3 = P - n_\alpha N_{\alpha\beta} u_{3,\beta}. \tag{8}$$

The constitutive equations were used to obtain t_α ($t_\alpha = M_{\alpha\beta}.n_\beta$) and t_3 ($t_3 = Q_\alpha.n_\alpha$), respectively, in the natural conditions.

The general form of the displacement boundary integral equations (DBIEs) with an additional domain integral containing the GNL effect is written next with the notation proposed by Weeën:

$$\frac{1}{2}C_{ij}(x')u_j(x') + \int_\Gamma \left[T_{ij}(x',x)u_j(x) - U_{ij}(x',x)t_j(x)\right]d\Gamma(x) = \cdots$$

$$= \iint_\Omega \left\{U_{i3}(x',X)\left[\frac{\partial}{\partial X_\alpha}\left(N_{\alpha\beta}\frac{\partial u_3}{\partial X_\beta}\right)\right] + U_{i\gamma}(x',X)\frac{h^2}{12}\left[\frac{\partial}{\partial X_\alpha}\left(N_{\alpha\beta}\frac{\partial u_\gamma}{\partial X_\beta}\right)\right]\right\}d\Omega(X). \qquad (9)$$

in which C_{ij} is an element of the matrix C related to the boundary at the source point, which becomes the identity matrix when a smooth boundary is considered, U_{ij} represents the rotation ($j = 1, 2$) or the deflection ($j = 3$) due to a unit couple ($i = 1, 2$) or a unit point force ($i = 3$), respectively, T_{ij} represents the moment ($j = 1, 2$) or the shear ($j = 3$) due to a unit couple ($i = 1, 2$) or a unit point force ($i = 3$), respectively.

It is well known in nonlinear analyses of beams or plates [1], [15] that the natural condition is introduced for each generalized force t_i corresponding to the displacement not prescribed. According to eqns (7) and (8), the GNL effect should be introduced when the deflection and/or rotations is/are not prescribed on the boundary portion.

The term related to the GNL effect in eqns (5), (6) or (9) can be simplified with the equilibrium equations for in-plane forces ($N_{\alpha\beta,\alpha} = 0$). The second derivatives of the displacements (rotations or deflections) result from the simplification, as shown in several studies in the literature. The equilibrium equations for in-plane forces were not used here as done in [4] but an algebraic manipulation with the divergence theorem was done in the domain integral related to GNL effect in eqn (9), i.e.:

$$\iint_\Omega \left\{U_{i3}(x',X)\left[\frac{\partial}{\partial X_\alpha}\left(N_{\alpha\beta}\frac{\partial u_3}{\partial X_\beta}\right)\right] + U_{i\gamma}(x',X)\frac{h^2}{12}\left[\frac{\partial}{\partial X_\alpha}\left(N_{\alpha\beta}\frac{\partial u_\gamma}{\partial X_\beta}\right)\right]\right\}d\Omega(X) = \cdots$$

$$\int_\Gamma \left[U_{i3}(x',x)n_\alpha(x)N_{\alpha\beta}(x)u_{3,\beta}(x) + \frac{h^2}{12}U_{i\gamma}(x',x)n_\alpha(x)N_{\alpha\beta}(x)u_{\gamma,\beta}(x)\right]d\Gamma(x) + \cdots$$

$$- \iint_\Omega \left[U_{i3,\alpha}(x',X)N_{\alpha\beta}(X)u_{3,\beta}(X) + \frac{h^2}{12}U_{i\gamma,\alpha}(x',X)N_{\alpha\beta}(X)u_{\gamma,\beta}(X)\right]d\Omega(X).$$

Two integrals containing the GNL effect result from the algebraic manipulation with the divergence theorem, where one is computed on the boundary and the other on the domain. Despite the increase in the number of integrals with the GNL effect, the first derivatives of the deflection and rotations were only necessary in two integrals and the equilibrium equations for in-plane forces were not required. The final DBIE is given by:

$$\frac{1}{2}C_{ij}(x')u_j(x') + \int_\Gamma \left[T_{ij}(x',x)u_j(x) - U_{ij}(x',x)t_j(x)\right]d\Gamma(x) = \cdots$$

$$= \int_\Gamma \left[U_{i3}(x',x)n_\alpha(x)N_{\alpha\beta}(x)u_{3,\beta}(x) + \frac{h^2}{12}U_{i\gamma}(x',x)n_\alpha(x)N_{\alpha\beta}(x)u_{\gamma,\beta}(x)\right]d\Gamma(x) + \cdots$$

$$- \iint_\Omega \left[U_{i3,\alpha}(x',X)N_{\alpha\beta}(X)u_{3,\beta}(X) + \frac{h^2}{12}U_{i\gamma,\alpha}(x',X)N_{\alpha\beta}(X)u_{\gamma,\beta}(X)\right]d\Omega(X). \tag{10}$$

The boundary integral containing the GNL effect can be related to natural conditions given by eqns (7) and (8) when the boundary portion has the deflection and rotations not prescribed. This can be shown by assuming the boundary Γ split into two portions: Γ_p and Γ_f where displacements are known (prescribed) and unknown (not prescribed or free), respectively.

$$\frac{1}{2}C_{ij}u_j + \int_{\Gamma_f} T_{ij}u_j\,d\Gamma - \int_{\Gamma_p} U_{ij}t_j\,d\Gamma = \int_{\Gamma_f} U_{i\alpha}\,EM_\alpha\,d\Gamma + \int_{\Gamma_f} U_{i3}\,P\,d\Gamma + \cdots$$

$$- \int_{\Gamma_p} T_{ij}u_j\,d\Gamma - \iint_\Omega \left[U_{i3,\alpha}N_{\alpha\beta}u_{3,\beta} + \frac{h^2}{12}U_{i\gamma,\alpha}N_{\alpha\beta}u_{\gamma,\beta}\right]d\Omega + \cdots$$

$$+ \int_{\Gamma_p} \left[U_{i3}n_\alpha N_{\alpha\beta}u_{3,\beta} + \frac{h^2}{12}U_{i\gamma}n_\alpha N_{\alpha\beta}u_{\gamma,\beta}\right]d\Gamma. \tag{11}$$

The left-hand side of eqn (11) contains the unknowns, i.e., displacements on Γ_f and forces on Γ_p. The loads on the boundary portion Γ_f were introduced in the right hand side according to natural conditions shown in eqns (7) and (8). A simplification can be done on the boundary portion Γ_f due to opposite signals of the natural condition and the boundary integral with the GNL effect, which only needs to be computed on the boundary portion with prescribed displacements (Γ_p) as the result.

The gradient of displacements is required in the DBIE for the buckling problem (eqn 10) to introduce the GNL effect. The BIE for the gradient of displacements at an internal point is obtained by differentiating the eqn (10) with respect to the coordinates of the source point (X'). The result is next written in terms of differentiation of the field point coordinates and using the tangential differential operator [16].

$$u_{i,\gamma}(X') = \int_\Gamma \left\{M_{i\alpha\beta}(X',x)D_{\gamma\alpha}[u_\beta(x)] + n_\gamma(x)Q_{i\beta}(X',x)u_\beta(x)\right\}d\Gamma(x) + \cdots$$

$$+ \int_\Gamma \left\{Q_{i\beta}(X',x)D_{\gamma\beta}[u_3(x)] - U_{ij,\gamma}(X',x)t_j(x)\right\}d\Gamma(x) + \cdots$$

$$+ \int_{\Gamma} \left[U_{i3,\gamma}(x',x)n_\alpha(x)N_{\alpha\beta}(x)u_{3,\beta}(x) + \frac{h^2}{12} U_{i\rho,\gamma}(x',x)n_\alpha(x)N_{\alpha\beta}(x)u_{\rho,\beta}(x) \right] d\Gamma(x) + \cdots$$

$$- \iint_{\Omega} \left[U_{i3,\alpha\gamma}(x',X)N_{\alpha\beta}(X)u_{3,\beta}(X) + \frac{h^2}{12} U_{i\rho,\alpha\gamma}(x',X)N_{\alpha\beta}(X)u_{\rho,\beta}(X) \right] d\Omega(X), \qquad (12)$$

with

$$D_{\alpha\beta}[f(x)] = n_\alpha(x)f_{,\beta}(x) - n_\beta(x)f_{,\alpha}(x).$$

3 THE NUMERICAL IMPLEMENTATION

In the formulation described in this paper, quadratic shape functions for isoparametric boundary elements were employed with collocation points always placed on the boundary. The same mapping function was used for conformal and non-conformal interpolations, i.e., nodes at ends of quadratic elements remain at ends when discontinuous elements were employed. The collocation points were placed at nodes in case of continuous elements and at positions (−0.67, 0.0, +0.67), in the range (−1, 1), in case of discontinuous elements, i.e., the collocation points were shifted inside the element at the corresponding end where the discontinuity exists. Singularity subtraction [17] and the transformation of variable technique [18] were employed for the Cauchy and the weak type of singularity, respectively, when integrations were performed on elements containing the collocation points. The standard Gauss-Legendre scheme was employed for integrations on elements (or side of the cell) not containing the collocation points. Rectangular cells were used to discretize the domain integral related to the GNL effect. The derivatives of the displacements (deflection and rotation) at the center of the cell were assumed constant on the cell. This assumption allowed the use of the divergence theorem to convert the domain integral into equivalent boundary integrals performed on sides of the cell. This strategy led to a simplification on the use of integrals containing the GNL effect because they have opposite signals as done in [4]. The GNL effect was computed from integrations performed on sides of cells inside the domain in all problems. The boundary condition is required to include the integrals with the GNL effect computed on sides of the cell on the plate boundary:

1. When the displacements are prescribed on the whole boundary (like a clamped plate on all sides), the GNL effect was not computed from integrations performed on sides of cells on the plate boundary.
2. When the displacements are not prescribed on the boundary portion of the plate (Γ_f), the GNL effect was computed from integrations performed on sides of cells on the boundary portion Γ_f.
3. In the case of a hard or soft boundary condition, when the deflection is prescribed and the tangential rotation or both rotations is/are not prescribed, the corresponding GNL effect related to curvatures is/are computed on the boundary portion related to the hard or soft condition.

The basic inverse iteration and the Rayleigh quotient were used to perform the eigenvalue analysis [19], i.e.:

$$Ax^{(k+1)} = Bx^k, \tag{13}$$

$$\lambda_k = \frac{(x^{(k+1)}, x^k)}{(x^{(k+1)}, x^{(k+1)})}. \tag{14}$$

The basic inverse iteration procedure is very efficient to compute the lower eigenvalues with corresponding eigenvectors [19]. The discretized forms of eqns (10) and (12) were used instead of eqn (13) as done in [4]. Starting with an eigenvector x^1 with elements equal to 1.0, the values of the displacements and tractions at the boundary nodes are found with eqn (10); these values are introduced in eqn (12) to obtain the gradient of the displacements (elements of the eigenvector x^2), and the lowest eigenvalue at the first iteration step was obtained by using eqn (14). The iteration procedure continued until the absolute difference between values of successive eigenvalues was less than 10^{-8}. The proof of convergence for the lower eigenvalues can be found in [19].

4 NUMERICAL EXAMPLES

The Young modulus (E) was 206.9 Gpa, the Poisson ratio (v) was 0.3. The buckling parameter k is a non-dimensional value related to the critical load of the plate (N_{cr}), the length of the plate side (a) and the flexural rigidity (D), which is obtained according to following expression:

$$k = \frac{a^2 N_{cr}}{\pi^2 D}.$$

The buckling parameter k was obtained according to the following boundary conditions: S = simply supported edge, C = clamped edge and F = free edge.

The results obtained were compared to those presented by Mizusawa [9], which were not different to those presented by Dawe and Roufaeil in [5], but more types of boundary conditions were studied in [9]. The Spline Strip Method was used in [9] and the simply supported boundary condition employed the hard restraint condition (tangential rotation is restrained). Results presented in Tables 1 and 2 used 128 quadratic boundary elements (260 nodes) and 256 constant cells as done in [4].

The differences in the buckling parameter obtained with the GNL effect using only deflection derivatives to those obtained with the present formulation were included in the last row of Tables 1 and 2. The results agreed with the comment on the effect of boundary condition done by Mizusawa in [9]. Results in Table 2 were not significantly changed when the curvatures were included in the GNL effect even for the highest thickness when the maximum difference to [4] was 5.02%. On the other hand, results with curvatures included in Table 1 were significant for higher thicknesses, as shown by Sun on buckling analyses for Timoshenko's beam in [7].

5 CONCLUSIONS

Results obtained for buckling analyses with BEM and including curvatures in the GNL effect agreed to those in the literature. The effect of curvatures was greater for the highest thickness in the BEM formulation with reference to those in [9]. The first derivatives of displacements (deflection and rotations) were used in the present formulation with the same algebraic manipulation done in [4] where only deflection derivatives were employed. Furthermore, the BIE for the gradient of displacements employed the tangential differential operator to reduce

Table 1: Buckling parameter (k) of the first critical load of square plates under uniaxial in-plane loading. Effect of curvatures relevant.

Type	h/a	[4]	[9]	This work	Diff. to [9] (%)	Diff. to [4] (%)
1) SSSS	0.001	4.0128	4.0	4.0127	0.32	0.00
	0.010	4.0105	3.997	4.0088	0.29	−0.04
	0.050	3.9561	3.928	3.9174	−0.27	−0.99
	0.100	3.7952	3.729	3.6638	**−1.78**	**−3.59**
	0.200	3.2643	3.119	2.9587	**−5.42**	**−10.33**
2) SSSC	0.001	4.8707	4.8470	4.8707	0.49	0.00
	0.010	4.8666	4.8420	4.8633	0.44	−0.07
	0.050	4.7681	4.7170	4.6940	−0.49	−1.58
	0.100	4.4857	4.3720	4.2541	**−2.77**	**−5.44**
	0.200	3.6251	3.4180	3.1927	**−7.06**	**−13.54**
3) CSSS	0.001	5.7597	5.7400	5.7597	0.34	0.00
	0.010	5.7538	5.7330	5.7509	0.31	−0.05
	0.050	5.6164	5.5740	5.5488	−0.45	−1.22
	0.100	5.2334	5.14	5.0205	**−2.38**	**−4.24**
	0.200	4.1468	3.8760	3.6066	**−7.47**	**−14.98**
4) SCSC	0.001	6.7967	6.7430	6.7966	0.79	0.00
	0.010	6.7875	6.7310	6.7787	0.70	−0.13
	0.050	6.5742	6.4620	6.3827	−1.24	−3.00
	0.100	5.9914	5.7650	5.4702	**−5.39**	**−9.53**
	0.200	4.4260	4.1090	3.7064	**−10.86**	**−19.42**
5) CSCS	0.001	7.7540	7.692	7.7539	0.80	0.00
	0.010	7.7370	7.671	7.7284	0.74	−0.11
	0.050	7.3559	7.228	7.1742	−0.75	−2.53
	0.100	6.4138	6.178	5.9525	**−3.79**	**−7.75**
	0.200	4.3413	4.056	3.7733	**−7.49**	**−15.05**
6) CCCC	0.001	10.1605		10.1603		0.00
	0.010	10.1382	10.055	10.1225	0.67	−0.16
	0.050	9.6326		9.3115		−3.45
	0.100	8.3374	8.047	7.5670	**−6.34**	**−10.18**
	0.200	5.3121		4.5041		**−17.94**

Table 2: Buckling parameter (k) of the first critical load of square plates under uniaxial in-plane loading. Effect of curvatures not relevant.

Type	h/a	[4]	[9]	This work	Diff. to [9] (%)	Diff. to [4] (%)
1) FSSS	0.001	1.4038	1.402	1.4006	−0.10	−0.23
	0.010	1.4029	1.400	1.4027	0.19	−0.01
	0.050	1.3849	1.378	1.3796	0.12	−0.38
	0.100	1.3442	1.327	1.3248	−0.17	−1.46
	0.200	1.2168	1.173	1.1601	**−1.11**	**−4.89**
2) SSSF	0.001	2.3623	2.366	2.3623	−0.16	0.00
	0.010	2.3529	2.353	2.3524	−0.03	−0.02
	0.050	2.2520	2.237	2.2403	0.15	−0.52
	0.100	2.0908	2.060	2.0544	−0.27	−1.77
	0.200	1.7178	1.657	1.6387	**−1.12**	**−4.83**
3) FSCS	0.001	1.6515	1.652	1.6515	−0.03	0.00
	0.010	1.6536	1.650	1.6533	0.20	−0.02
	0.050	1.6246	1.615	1.6180	0.19	−0.41
	0.100	1.5604	1.539	1.5369	−0.14	−1.53
	0.200	1.3738	1.323	1.3085	**−1.11**	**−4.99**
4) SCSF	0.001	2.3879	2.392	2.3879	−0.17	0.00
	0.010	2.3787	2.378	2.3782	0.01	−0.02
	0.050	2.2747	2.260	2.2624	0.11	−0.54
	0.100	2.1090	2.078	2.0708	−0.35	−1.84
	0.200	1.7274	1.666	1.6448	**−1.29**	**−5.02**
5) FSFS	0.001	0.9505	0.9523	0.9505	−0.19	0.00
	0.010	0.9533	0.9516	0.9532	0.17	−0.01
	0.050	0.9449	0.9412	0.9423	0.12	−0.28
	0.100	0.9236	0.9146	0.9137	−0.10	−1.08
	0.200	0.8516	0.8274	0.8213	**−0.74**	**−3.69**
6) SFSF	0.001	2.0370	2.043	2.0370	−0.29	0.00
	0.010	2.0308	2.032	2.0305	−0.07	−0.01
	0.050	1.9508	1.942	1.9449	0.15	−0.30
	0.100	1.8271	1.807	1.8083	0.07	−1.04
	0.200	1.5389	1.497	1.4949	**−0.14**	**−2.94**

the order of singularity in the kernels of integrals. A BEM formulation for buckling analyses employing only the first derivatives of displacements and including a BIE for the gradient of displacement with singularities reduced were the main numerical features of the present formulation, which present results closer to the literature with a low number of degrees of freedom related to the boundary element analysis.

ACKNOWLEDGEMENT

The authors are grateful to CAPES (88882.435160/2019-01) for the support in the development of this research.

REFERENCES

[1]   Timoshenko, S.P. & Woinowsky-Krieger, S., *Theory of Plates and Shells*, 2nd Ed., McGraw-Hill Book Company: New York, 1959.
[2]   Reissner, E., The effect of transverse shear deformation on the bending of elastic plates. *Journal of Applied Mechanics*, **12**(2), pp. A66–A77, 1945.
[3]   Mindlin, R.D., Influence of rotatory inertia and shear on flexural motions of isotropic elastic plates. *Journal of Applied Mechanics*, **18**, pp. 18–31, 1951.
[4]   Soares, R.A., Jr. & Palermo, L., Jr., Effect of shear deformation on the buckling parameter of perforated and non-perforated plates studied using the boundary element method. *Engineering Analysis with Boundary Elements*, **85**, pp. 57–69, 2017.
[5]   Dawe, D.J. & Roufaeil, O.L., Buckling of rectangular Mindlin plates. *Composite Structures*, **15**(4), pp. 461–471, 1982.
[6]   Herrmann, G. & Armenakas, A.E., Vibration and stability of plates under initial stress. *Transactions of the American Society of Civil Engineers*, **127**, pp. 458–487, 1962.
[7]   Sun, C.T., On the equations for a Timoshenko beam under initial stress. *Journal of Applied Mechanics*, 39, pp. 282–285, 1972.
[8]   Brunelle, E.J. & Robertson, S.R., Initially stressed Mindlin plates. *AIAA Journal*, **12**, pp. 1036–1045, 1974.
[9]   Mizusawa, T., Buckling of rectangular Mindlin plates with tapered thickness by the spline strip method. *International Journal of Solids & Structures*, **30**(12), pp. 1663–1677, 1993.
[10]   Smith, J.P., Buckling of shear deformable plates using the p-version of the finite element method. *Composite Structures*, **57**(3), pp. 527–532, 1995.
[11]   Doong, J.L., Vibration and stability of an initially stressed thick plate according to a high-order deformation theory. *Journal of Sound and Vibration*, **113**, pp. 425–440.
[12]   Matsunaga, H., Free vibration and stability of thick elastic plates subjected to in-plane forces. *International Journal of Solids and Structures*, **31**(22), pp. 3113–3124, 1994.
[13]   Elsgoltz, L., *Ecuaciones Diferenciales y Cálculo Variacional*, Editorial Mir: Moscou, 1977.
[14]   Katsikadelis, J.T., *The Boundary Element Method for Plate Analysis*, Elsevier Academic Press, 2014.
[15]   Timoshenko, S.P. & Gere, J.M., *Theory of Elastic Stability*, 2nd Ed., Dover Publication: New York, 1961.
[16]   Palermo L., Jr., The tangential differential operator applied to a stress boundary integral equation for plate bending including the shear deformation effect. *Engineering Analysis with Boundary Elements*, **36**, pp. 1213–1225, 2012.
[17]   Wrobel, L.C., *The Boundary Element Method: Applications in Thermo-Fluids and Acoustics*, John Wiley & Sons Ltd.: Chichester, 2002.

[18]   Telles, J.C.F., A self-adaptive coordinate transformation for efficient numerical evaluation of general boundary element integrals. *International Journal for Numerical Methods in Engineering*, **24**, pp. 959–973, 1987.

[19]   Wilkinson, J.H., *The Algebraic Eigenvalue Problem*, Oxford University Press, William Clowes & Sons: London, 1972.

MITIGATION OF WAVE LOADS ON THE FLOATING POROUS STRUCTURE BY SLOTTED SCREENS

KOTTALA PANDURANGA & SANTANU KOLEY
Department of Mathematics, Birla Institute of Technology and Science – Pilani, Hyderabad Campus, India

ABSTRACT

This paper presents the effectiveness of partially immersed slotted screens placed at a finite distance away from the floating porous breakwater to mitigate the wave-induced hydrodynamic forces on the floating breakwater. The fluid motion within the porous structure is analyzed using the Sollitt and Cross model. Further, a non-linear pressure drop condition on the slotted screens is considered, which includes the effects of wave height on the incident wave energy dissipation by the slotted screens. In the presence of the non-linear boundary condition on the slotted screen, the associated boundary value problem is solved numerically using the iterative multi-domain boundary element method. Various physical interests of the study, such as reflection coefficient, transmission coefficient, wave energy dissipation coefficient, and wave forces acting on the slotted screens and floating breakwater, are analyzed in detail for various wave and structural parameters. The results show that the maximum wave-induced forces acting on the porous breakwater are observed when the gap between the slotted screen and porous breakwater is approximately equal to an integer multiple of half of the wavelength for different porosities of the slotted screens.

Keywords: slotted screens, wave-induced forces, iterative multi-domain boundary element method, reflection coefficient.

1 INTRODUCTION

Floating porous structures are commonly used as effective wave energy attenuators in the coastal environment to protect the harbours and seawall structures from wave attacks. Due to the continuous wave action on the floating porous structures and the demand of the serviceability requirements, it is necessary to mitigate the wave-induced hydrodynamic forces acting on the structures.

There are several ways to mitigate the wave-induced forces and structural responses of floating structures. One among them is the installation of the bottom-founded breakwater (Wang et al. [1]). However, the building of such conventional breakwaters is inadequate in reducing the structural response of floating structures as they prevent the water circulation around the structure, leading to environmental damage (Hong et al. [2]). Moreover, the construction of the conventional breakwaters is complex and more expensive as the depth of the sea increases. Therefore, several researchers proposed alternative wave attenuating structures that could effectively mitigate the wave forces and work as an anti-motion device for the floating structure such as the floating breakwater (Tay et al. [3]), mooring lines (Nguyen et al. [4]), horizontal and vertical anti-motion plates attached to the structures (Ohta [5]), submerged horizontal and inclined porous plates (Watanabe et al. [6], Cheng et al. [7]), gill cells with the perforated bottom surface (Wang et al. [8]), etc. A semi-rigid connector in reducing the structural response of the interconnected beams was used by Wang et al. [9]. Further, the study was extended by Gao et al. [10] using a flexible hinge-line connector. It was observed that a flexible hinge-line connector was found to be more effective in reducing the structural deflection of the floating VLFS than a semi-rigid connector when the incident wavelength is small. Further, installing an anti-motion device such as OWC-WEC (oscillating water column wave energy converter device) in front of the floating VLFS can also effectively decreases the hydroelastic responses of floating VLFSs. Hong et al. [11]

WIT Transactions on Engineering Sciences, Vol 131, © 2021 WIT Press
www.witpress.com, ISSN 1743-3533 (on-line)
doi:10.2495/BE440091

proposed an analytical method to evaluate the structural reduction efficiency of a T-shaped freely floating breakwater with a built-in OWC device.

The aforementioned anti-motion systems can also be used to reduce wave-induced hydrodynamic forces on the floating breakwater, in addition to mitigating wave-induced structural responses. Huang et al. [12] presented a detailed review on the wave reflection and transmission characteristics of perforated/slotted screens, as well as the incident wave energy dissipation mechanism by perforated/slotted vertical and horizontal screens, and the reduction of wave-induced forces acting on the bottom standing structures by surface-piercing floating perforated/slotted screens. Vijay et al. [13] studied the effectiveness of pair of permeable plates on reducing structural responses and wave forces on the floating structure using the multi-domain boundary element method. Their study revealed that the floating structure experiences the minimum wave forces when the twin porous plates are positioned in the middle of the floating structure and the rigid wall. Sun et al. [14] examined the installation of a submerged porous breakwater to mitigate wave loads on the floating bridge decks. It was seen that the construction of the submerged porous breakwater at suitable locations could significantly reduce the horizontal wave loads on the floating structure. Using the method of matched eigenfunction expansions, Qiao et al. [15] studied the motion responses and wave forces on the floating rigid breakwater attached to a pair of perforated side plates. The wave-induced hydrodynamic forces on the floating rigid body reduced significantly with an increase in the porous-effect parameter of the sides-walls when the breakwater system is rotated about the center (lies at the bottom) of the floating body. From these studies, it is observed that the slotted screens have a great impact on mitigating the wave-induced loads on floating bodies. Further, several researchers considered the linearized pressure drop condition across the perforated/slotted breakwater using the Darcy's law (Qiao et al. [15], Yip and Chwang [16]). But in general, the flow separation through the slotted screen is quadratic in nature. The significance of this quadratic pressure drop condition is that it includes the effects of wave height in dissipating the incident wave energy (Liu and Li [17]).

In the present work, the mitigation of wave forces on the rectangular-shaped floating porous breakwater by placing a pair of slotted screens on either side of the floating porous breakwater is studied using the linear potential flow theory. The fluid motion in the porous breakwater is analyzed using the Sollitt and Cross model [18]. Further, a non-linear (quadratic) pressure drop condition on the slotted screen is adopted to study the efficiency of the slotted screens on mitigating the wave loads acting on the floating porous breakwater. An iterative boundary element method (BEM) is adopted to handle the non-linear pressure drop condition on the slotted screen.

2 MATHEMATICAL FORMULATION

The schematic geometry of the rectangular-shaped floating porous breakwater protected by two slotted screens placed on either side of the breakwater is shown in Fig. 1. To study the interaction of the regular waves with the floating porous breakwater system, a two-dimensional Cartesian system is adopted where the x-axis is taken as the horizontal axis, and z-axis is taken vertically upward. Consider linear small-amplitude simple harmonic water waves of wave height H, wavelength L, and circular frequency ω propagating along with the positive x-axis. For mathematical description, the structural parameters are represented as water depth h, submergence depth of the porous breakwater d_2, the width of the porous breakwater B, submergence depth of the slotted screens d_1. The space between the slotted screen and breakwater is w (on either side of the breakwater). In the presence of the breakwater system and pair of slotted screens, the total fluid domain is divided into six regions Ω_j, $j = 1, 2, \ldots 6$, as shown in Fig. 1. The computational domain of the present physical problem is shown in Fig. 2.

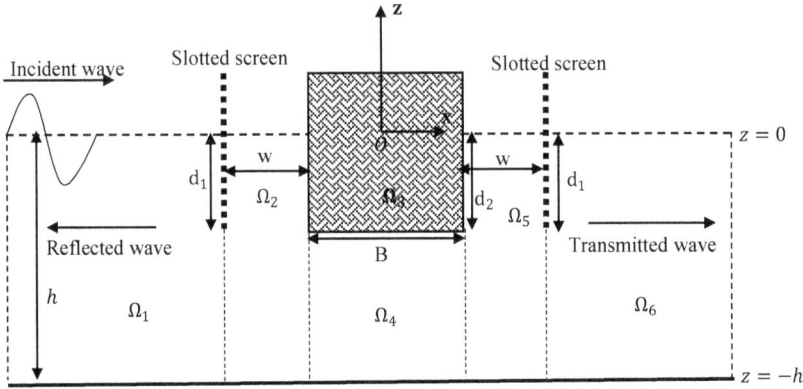

Figure 1:   Schematic diagram of floating porous breakwater protected by pair of slotted screens.

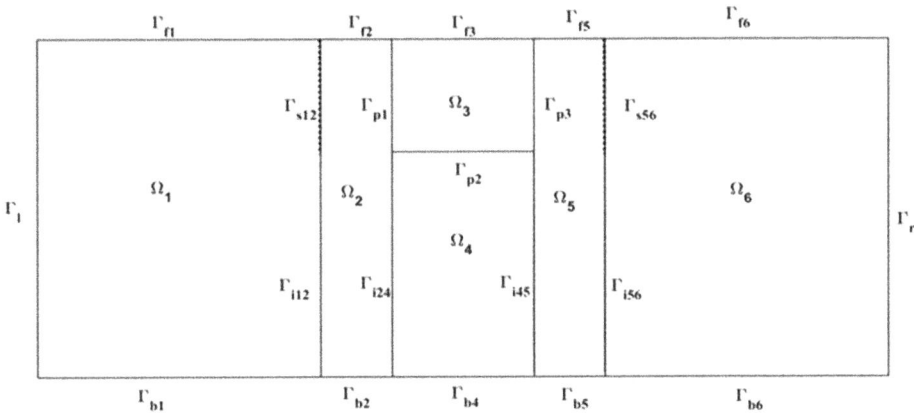

Figure 2: Computational domain of the physical problem.

It is assumed that the linear potential flow theory applies, i.e., the flow is incompressible, inviscid, and its motion is irrotational. Then, the velocity potential $\Phi_j(x, z, t)$ in every region is expressed as $\Phi(x, z, t) = \text{Re}\{\phi(x, z)e^{-i\omega t}\}$, where the spatial velocity potentials satisfy the Laplace equation

$$\frac{\partial^2 \phi^{(j)}}{\partial x^2} + \frac{\partial^2 \phi^{(j)}}{\partial z^2} = 0, \quad j = 1, 2, \ldots 6, \tag{1}$$

where the superscript j corresponds to the regions j for $j = 1, 2, \ldots 6$.

The velocity potentials $\phi^{(j)}$ also satisfy the following boundary conditions (BC)

1. The linearized free-surface BC at $z = 0$ is given by

$$\frac{\partial \phi^{(j)}}{\partial z} - \frac{\omega^2}{g}(s + if)\,\phi^{(j)} = 0, \quad \text{on} \quad \Gamma_{fj} \text{ for } j = 1, 2, 3, 5, 6, \tag{2}$$

where s is the inertial coefficient and f is the friction coefficient of the porous region. It is to be noted that $s = 1, f = 0$ should be taken for the open water regions.

2. The interface BCs between the regions $\Omega_1, \Omega_2, \Omega_4, \Omega_5,$ and Ω_6 are written as

$$\phi^{(1)} = \phi^{(2)}, \quad \frac{\partial \phi^{(1)}}{\partial x} = \frac{\partial \phi^{(2)}}{\partial x}, \quad \text{on } \Gamma_{i12}, \quad \phi^{(2)} = \phi^{(4)}, \quad \frac{\partial \phi^{(2)}}{\partial x} = \frac{\partial \phi^{(4)}}{\partial x}, \quad \text{on } \Gamma_{i24}, \tag{3}$$

$$\phi^{(4)} = \phi^{(5)}, \quad \frac{\partial \phi^{(4)}}{\partial x} = \frac{\partial \phi^{(5)}}{\partial x}, \quad \text{on } \Gamma_{i45}, \quad \phi^{(5)} = \phi^{(6)}, \quad \frac{\partial \phi^{(5)}}{\partial x} = \frac{\partial \phi^{(6)}}{\partial x}, \quad \text{on } \Gamma_{i56}. \tag{4}$$

3. The BC on the horizontal seabed is written as

$$\frac{\partial \phi^{(j)}}{\partial z} = 0, \quad \text{on} \quad \Gamma_{bj} \text{ for } j = 1, 2, 4, 5, 6. \tag{5}$$

4. The interface BCs between the open water regions Ω_2, Ω_5, and porous regions Ω_3, Ω_4 are given by [18]

$$\begin{cases} \phi^{(2)} = (s + if)\,\phi^{(3)}, \quad \dfrac{\partial \phi^{(2)}}{\partial x} = \epsilon \dfrac{\partial \phi^{(3)}}{\partial x}, \text{ on } \Gamma_{p1}, \\[2mm] \phi^{(4)} = (s + if)\,\phi^{(3)}, \quad \dfrac{\partial \phi^{(4)}}{\partial z} = \epsilon \dfrac{\partial \phi^{(3)}}{\partial z}, \text{ on } \Gamma_{p2}, \\[2mm] \phi^{(5)} = (s + if)\,\phi^{(3)}, \quad \dfrac{\partial \phi^{(5)}}{\partial x} = \epsilon \dfrac{\partial \phi^{(3)}}{\partial x}, \text{ on } \Gamma_{p3}. \end{cases} \tag{6}$$

5. The non-linear (quadratic) pressure drop condition and the continuity of horizontal fluid velocity at the slotted screen is given by [17]

$$\begin{cases} \phi^{(1)} - \phi^{(2)} = -\dfrac{8i}{3\pi\omega}\dfrac{1-\tau_1}{2\mu\tau_1^2}\left|\dfrac{\partial \phi^{(1)}}{\partial x}\right|\dfrac{\partial \phi^{(1)}}{\partial x} - 2C_1 \dfrac{\partial \phi^{(1)}}{\partial x}, \\[2mm] \dfrac{\partial \phi^{(1)}}{\partial x} = \dfrac{\partial \phi^{(2)}}{\partial x}, \end{cases} \quad \text{on } \Gamma_{s12}, \tag{7}$$

$$\begin{cases} \phi^{(5)} - \phi^{(6)} = -\dfrac{8i}{3\pi\omega}\dfrac{1-\tau_2}{2\mu\tau_2^2}\left|\dfrac{\partial \phi^{(5)}}{\partial x}\right|\dfrac{\partial \phi^{(5)}}{\partial x} - 2C_2 \dfrac{\partial \phi^{(5)}}{\partial x}, \\[2mm] \dfrac{\partial \phi^{(5)}}{\partial x} = \dfrac{\partial \phi^{(6)}}{\partial x}, \end{cases} \quad \text{on } \Gamma_{s56}, \tag{8}$$

where τ_j ($j = 1, 2$) are the porosity (geometrical) of the slotted screens, and μ is the discharge coefficient, respectively. Further, the expressions for the blockage coefficients C_j are given by [17]

$$C_j = 0.5d \left(\frac{1}{\tau_j} - 1 \right) + \frac{\delta}{\pi} \left\{ 1 - \ln \tau_j + \frac{1}{3}\tau_j^2 + \frac{281}{180}\tau_j^4 \right\}, \qquad j = 1, 2,$$

with d being the thickness of the screen and δ being the distance between the centers of two adjacent slots.

6. The far-field BCs at the two auxiliary boundaries placed at $x = -l$ and $x = r$ are represented as

$$\lim_{x \to -\infty} \left(\frac{\partial(\phi^{(1)} - \phi^{(0)})}{\partial x} + ik_0(\phi^{(1)} - \phi^{(0)}) \right) = 0, \tag{9a}$$

$$\lim_{x \to \infty} \left(\frac{\partial \phi^{(6)}}{\partial x} - ik_0 \phi^{(6)} \right) = 0, \tag{9b}$$

where $\phi^{(0)} = -\frac{igH}{2\omega} e^{ik_0 x} Z_0(z)$, $Z_0(z) = \frac{\cosh k_0(z+h)}{\cosh k_0 h}$, and k_0 is the real positive root of the dispersion relation $\omega^2 = gk \tanh kh$.

3 ITERATIVE BOUNDARY ELEMENT METHOD

In this section, an iterative boundary element method is adopted to solve the associated boundary value problem, as discussed in Section 2. Applying Green's second identity to the velocity potential ϕ and Green's function G, we obtain the resulting integral equation in each of the regions Ω_j for $j = 1, 2, \ldots 6$, as ([20])

$$\frac{1}{2}\phi^{(j)}(\xi, \eta) = \int_\Gamma \left(\phi^{(j)} \frac{\partial G^{(j)}}{\partial n_j} - G^{(j)} \frac{\partial \phi^{(j)}}{\partial n_j} \right) d\Gamma, \tag{10}$$

where n_j represents the unit outward normal to the boundary Γ_j of region j, and the Green's function and its normal derivatives are given by

$$G^{(j)} = \frac{1}{2\pi} \ln r, \frac{\partial G^{(j)}}{\partial n} = \frac{1}{2\pi r} \frac{\partial r}{\partial n}, \quad r = \sqrt{(x - \xi)^2 + (z - \eta)^2}. \tag{11}$$

In eqn (11), (x, z) and (ξ, η) are the field and source points, respectively. Now, discretize the boundaries of each of the regions Ω_j, $j = 1, 2, \ldots, 6$, into N_j (number of boundary elements on the boundary Γ_j of the region Ω_j) smooth elements. It is assumed that on each boundary element, $\phi^{(j)}$ and $\partial \phi^{(j)}/\partial n$ are constants. The discretized form of eqn (10) can be written as [20], [21]

$$\left[\alpha_{mn}^{(j)} \right] \left\{ \phi_n^{(j)} \right\} + \left[\beta_{mn}^{(j)} \right] \left\{ \frac{\partial \phi_n^{(j)}}{\partial n_j} \right\} = 0, \tag{12}$$

for $m, n = 1, 2, 3, \dots, N_j$, $j = 1, 2, \dots, 6$, in which the influence coefficients $\alpha_{mn}^{(j)}$ and $\beta_{mn}^{(j)}$ are expressed as

$$\alpha_{mn}^{(j)} = -\frac{1}{2}\delta_{mn} + \int_{\Gamma_{j,n}} \frac{\partial G^{(j)}}{\partial n_j}\, d\Gamma, \quad \beta_{mn}^{(j)} = \int_{\Gamma_{j,n}} G^{(j)}\, d\Gamma. \tag{13}$$

In eqn (12), $\phi_n^{(j)}$ and $\frac{\partial \phi_n^{(j)}}{\partial n_j}$ denote the velocity potential and its normal derivative at the midpoint of the nth boundary element of jth region. Further, δ_{mn} represents the Kronecker delta. On the other hand, the slotted screens share the same elements, i.e., the first (left) slotted screen shares regions 1 and 2, similarly the second (right) slotted screen shares the regions 5 and 6. Therefore, on the first slotted screen, i.e., on Γ_{s12}, the mth boundary element of region 1 and n^{th} boundary element of region 2 is the same. The pressure drop condition on the boundary Γ_{s12} is discretized as [17]

$$\phi_n^{(2)} = \phi_m^{(1)} + Z_{1m}\frac{\partial \phi_m^{(1)}}{\partial n_1}, \tag{14}$$

where

$$Z_{1m} = \frac{8i}{3\pi\omega}\frac{1-\tau_1}{2\mu\tau_1^2}\left|\frac{\partial \phi_m^{(1)}}{\partial n_1}\right| + 2C_1. \tag{15}$$

Further, the discretized form of the continuity of horizontal velocity is written as

$$\frac{\partial \phi_n^{(2)}}{\partial n_2} = -\frac{\partial \phi_m^{(1)}}{\partial n_1}. \tag{16}$$

Similarly, the boundary conditions on the second slotted screen, i.e., on Γ_{s56} are discretized as

$$\frac{\partial \phi_n^{(6)}}{\partial n_6} = -\frac{\partial \phi_m^{(5)}}{\partial n_5}, \tag{17}$$

$$\phi_n^{(6)} = \phi_m^{(5)} + Z_{1m}\frac{\partial \phi_m^{(5)}}{\partial n_5}, \tag{18}$$

where

$$Z_{2m} = \frac{8i}{3\pi\omega}\frac{1-\tau_2}{2\mu\tau_2^2}\left|\frac{\partial \phi_m^{(5)}}{\partial n_5}\right| + 2C_2. \tag{19}$$

In a similar manner, all other boundary conditions are discretized. Now, substitute all the discretized BCs along with eqns (14)–(19) into eqn (12) to obtain $\phi^{(j)}$ and $\partial\phi^{(j)}/\partial n$ over all boundary elements of each of the regions. It is to be noted that eqns (14) and (18) are non-linear. Thus, an iterative procedure is needed to evaluate them. The steps for the iterative procedure are as follows.

Step 1: Set $\frac{\partial \phi_m^{(1)}}{\partial n_1}$ and $\frac{\partial \phi_m^{(5)}}{\partial n_5}$ equal to zero (initial guess) in eqns (14) and (18) for the initial iteration, and determine Z_{1m} and Z_{2m}. Then the algebraic system (12) becomes linear.

Step 2: Solve the algebraic system (12) using the Gauss elimination method to get the updated values of $\frac{\partial \phi_m^{(1)}}{\partial n_1}$ and $\frac{\partial \phi_m^{(5)}}{\partial n_5}$.

Step 3: If the difference between the initial and updated values of $\frac{\partial \phi_m^{(1)}}{\partial n_1}$ and $\frac{\partial \phi_m^{(5)}}{\partial n_5}$ lies within the required error limit (i.e., 10^{-4}), stop at step 2. Otherwise, take the average values updated $\frac{\partial \phi_m^{(1)}}{\partial n_1}$ and $\frac{\partial \phi_m^{(5)}}{\partial n_5}$ and set it as a new initial guess and repeat from Step 2. In the present analysis, the required accuracy is obtained for not more than 20 iterations. Once $\phi_n^{(j)}$ and $\frac{\partial \phi_n^{(j)}}{\partial n_j}$ are obtained over each boundary element, the reflection coefficient C_R and transmission coefficient C_T are obtained by

$$C_R = \left| -1 + \frac{1}{N_0^2} \frac{2i\omega}{gH} e^{ik_0 l} \int_{-h}^{0} \phi^{(1)}(-l,z) Z_0(z) dz \right|, \tag{20}$$

$$C_T = \left| \frac{1}{N_0^2} \frac{2i\omega}{gH} \int_{-h}^{0} \phi^{(6)}(r,z) Z_0(z) dz \right|, \tag{21}$$

where $N_0^2 = \int_{-h}^{0} Z_0^2(z) dz$. The energy loss coefficient C_L is defined as (see [22] for detailed derivations)

$$C_L = (1 - C_R^2 - C_T^2). \tag{22}$$

The dimensionless horizontal and vertical wave forces acting on the porous floating breakwater are evaluated using the following.

$$F_x = \left| \frac{\omega}{gh^2}(s+if) \int_{\Gamma_{p1}} \phi^{(3)} n_x \, d\Gamma \right|, \qquad F_z = \left| \frac{\omega}{gh^2}(s+if) \int_{\Gamma_{p2}} \phi^{(3)} n_z \, d\Gamma \right|. \tag{23}$$

4 VALIDATIONS

In Fig. 3(a) and 3(b), the present iterative-BEM-based solutions are compared with the analytical solutions of Hu et al. [23] and Zhu and Chwang [19]. In Fig. 3(a), the reflection coefficient C_R and the transmission coefficient C_T are plotted against the non-dimensional wave number $k_0 d_2$ with the physical parameters $d_2 = 1$ and $h = 6.0 d_2$ for a floating rigid structure without slotted screens. It is well known that the porous structure and slotted screens become rigid when $\epsilon = 0.0, s = 1.0, f = 0.0$ and $\tau_j = 0.0, j = 1,2$, respectively. On the other hand, the slotted screen will disappear when the porosity of the screen is set to unity. As a limiting case to compare the present numerical solutions with Zhu and Chwang's [19] theoretical solutions, the floating rigid structure, and the first slotted screen with $\tau_1 = 0.5$ are fully extended till the bottom seabed with $h = 8\ m$. The comparisons reveal that there a good agreement between the present iterative-BEM solutions and theoretical results.

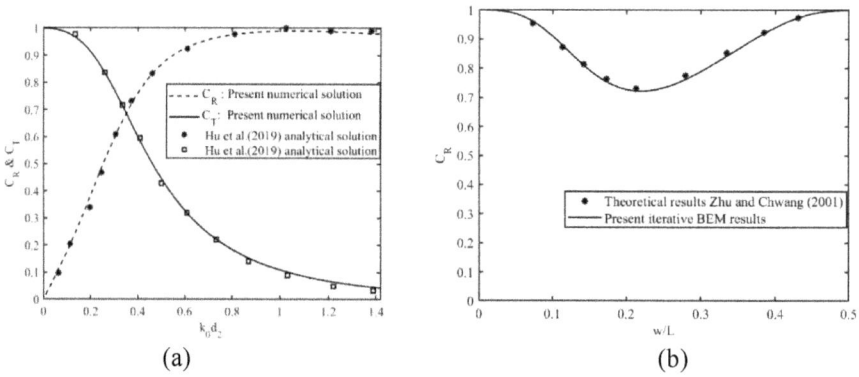

Figure 3:   Comparison of present numerical solution with (a) Hu et al. [23]; and (b) Zhu and Chwang [19] theoretical solutions.

5 RESULTS AND DISCUSSIONS

Here, some of the numerical results are presented to study the effectiveness of the slotted screens to mitigate the wave-induced forces acting on the floating porous breakwater. For the computational calculations, the physical parameters are considered as follows: $h = 10\ m$, $d_1 = \frac{h}{3}, d_2 = \frac{h}{5}$, $B = 2h\ w = \frac{h}{2}$, $s = 1.0, f = 1.0$, $\mu = 0.5, \epsilon = 0.1, C_j = h/10$, $\tau_1 = \tau_2 = 0.1$, unless otherwise mentioned.

In Fig. 4(a)–4(e), C_R, C_T, C_L, F_x, and F_z are plotted as a function of the non-dimensional relative space between the slotted screen and the porous breakwater w/L (L is the incident wavelength) for different porosities of the slotted screens. It is observed that C_R, C_T, C_L, F_x, and F_z follow a periodic oscillatory pattern as the gap between the slotted screen and the floating porous breakwater increases. This periodic nature is due to continuous reflection and transmission by the slotted screens and porous breakwater. Fig. 4(a) shows that C_R decreases with an increase in the porosity of the slotted screens. This is due to a large amount of incident wave energy being dissipated by the slotted screens and a porous floating breakwater, as seen in Fig. 4(c). Moreover, there is a leftward shift in the maximum of the reflection coefficient, and this is due to the phase shift of reflected waves by the slotted screens and the breakwater. In Fig. 4(b), C_T increases with an increase in the porosity of the screens. It is observed that more than 98% of the wave energy dissipated with 20% porosity of slotted screens and 10% porosity of the porous breakwater. Further, Fig. 4(d) and 4(e) shows that the non-dimensional horizontal force F_x and vertical force F_z decrease with an increase in the geometrical porosity of the slotted screen. The reason is that a significant portion of the incoming wave energy is dissipated by the breakwater and slotted screens.

In both Fig. 4(d) and 4(e), the maximum wave-induced forces are observed when $\frac{w}{l} \approx \frac{n}{2}$, $n = 1,2,3, ...$, and the minimum wave-induced forces acting on the porous breakwater are observed in the intermediate points of $\frac{(n-1)}{2} < \frac{w}{L} < \frac{n}{2}, n = 1,2,3, ...$ for different porosities of the slotted screens.

In Fig. 5(a)–5(e), C_R, C_T, C_L, F_x, and F_z are plotted as a function of the non-dimensional relative space between the slotted screen and the porous breakwater $\frac{w}{L}$ (L is the incident wavelength) for different submergence draft of the slotted screens $\frac{d_1}{h}$ with $\tau_1 = \tau_2 = 0.1$. It

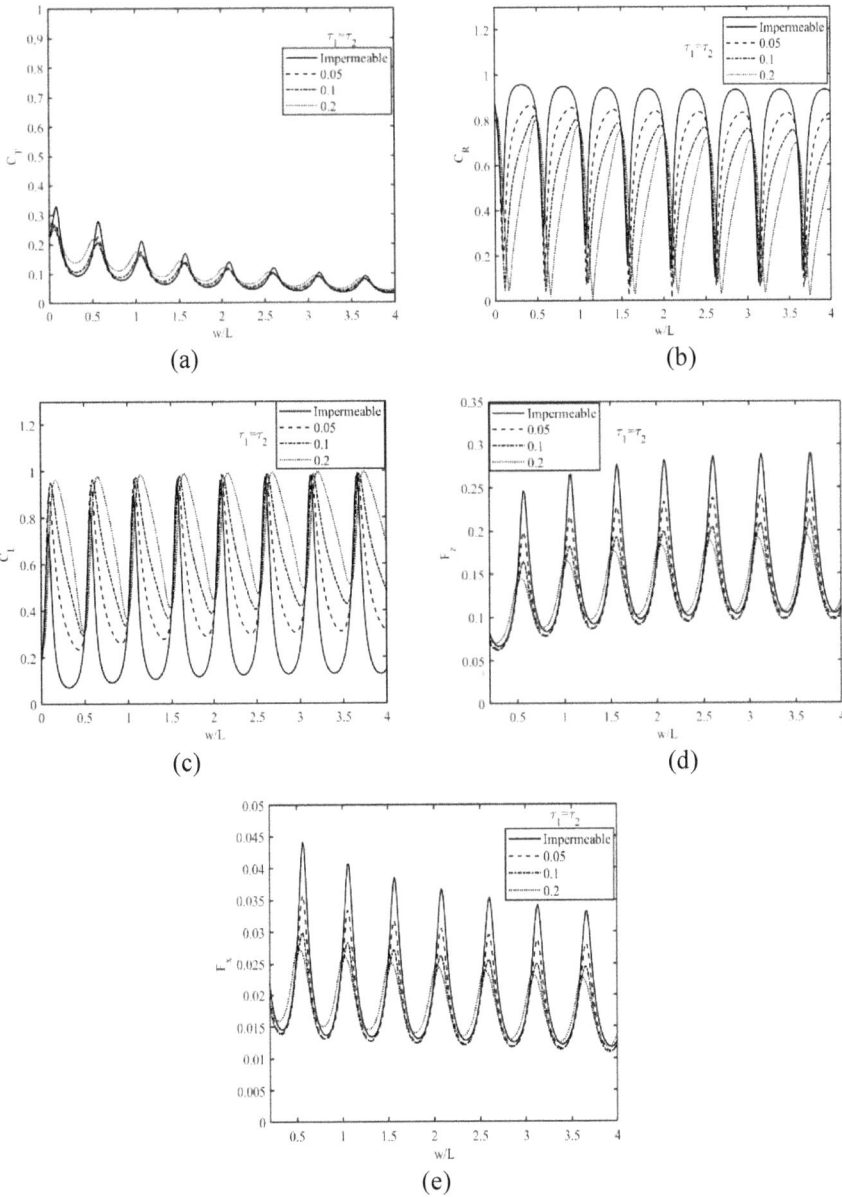

Figure 4: Variations of (a) C_R; (b) C_T; (c) C_L; (d) F_x; and (e) F_z vs w/L for different porosity of the slotted screens with $\tau_1 = \tau_2$, $k_0 h = 2.0$.

is observed that C_R, C_T, C_L, F_x, and F_z follow a periodic oscillatory pattern as the gap between the slotted screen and the floating porous breakwater increases. It is seen that in Fig. 5(a) and 5(b), the optimum C_R and C_T depend on the submergence draft of the surface-piercing slotted screens. However, the position of optimum does not depend on the submergence draft of the

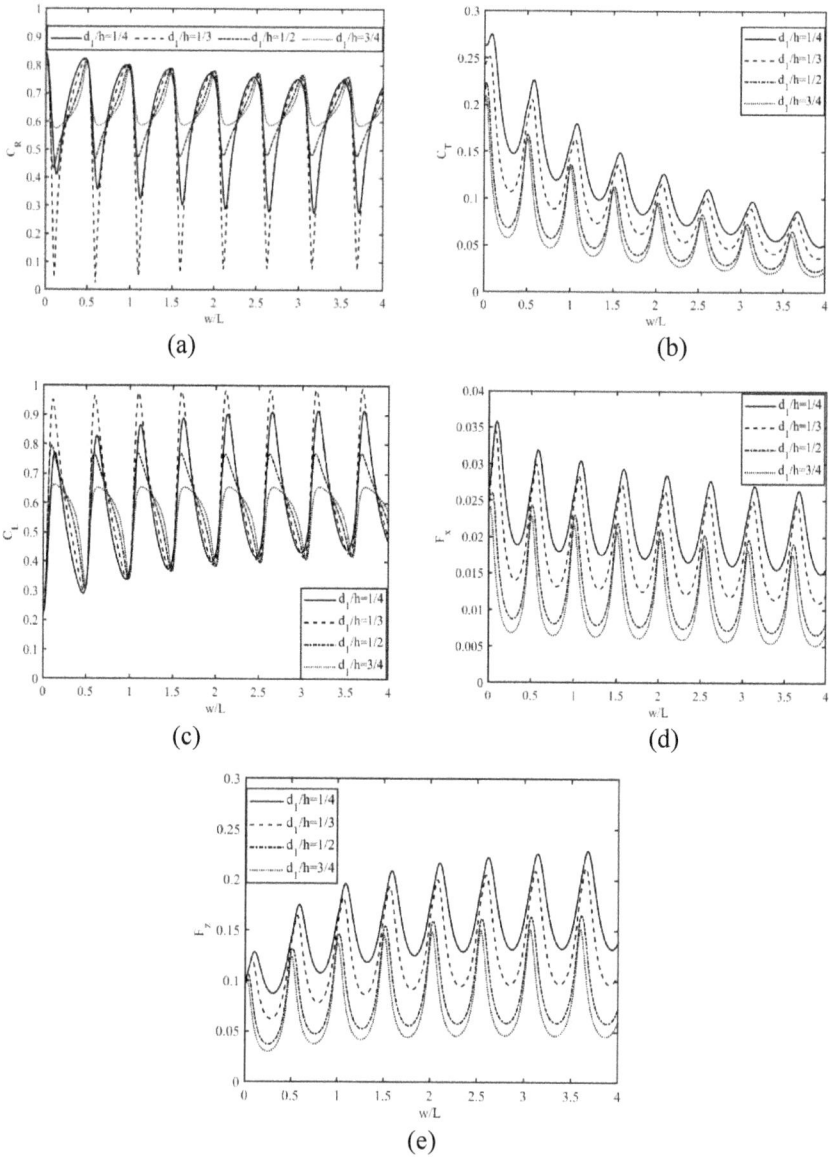

Figure 5:   Variations of (a) C_R; (b) C_T; (c) C_L; (d) F_x; and (e) F_z vs w/L for different submergence draft of the slotted screens $\frac{d_1}{h}$ with $\tau_1 = \tau_2 = 0.1$, $k_0 h = 2.0$.

surface-piercing slotted screens. In general, more amount of wave energy is concentrated at the free-surface. Therefore, surface-piercing slotted screens of shorter lengths dissipate more amount of wave energy. It is seen that for the present study, a slotted screen of submergence draft $\frac{d_1}{h} = \frac{1}{3}$ dissipates the maximum amount of incident wave energy with 10% screen porosity. The maximum and minimum wave-induced horizontal and vertical forces on the floating breakwater depend on the submergence draft of the slotted screen. However, the position of optimum does not depend on the submergence draft of the surface-piercing slotted screen. It is observed that the optimum positions are the same as that of Fig. 4(d) and 4(e) with small rightward phase shifts in the maximum wave forces.

6 CONCLUSIONS

This study develops an iterative-BEM-based numerical solution to study the effectiveness of slotted screens to mitigate the wave-induced hydrodynamic forces acting on the floating porous breakwater. The physical problem is studied within the framework of linear water wave theory. Further, the numerical results are validated with previous works as a limiting case of the present model. The numerical results show that

- The reflection, transmission, wave energy dissipation coefficients, and horizontal and vertical wave forces on the floating structure follow a periodic oscillatory pattern as the gap between the slotted screen and the floating porous breakwater increases.
- The reflection coefficient decreases with an increase in the porosity of the slotted screen. Further, more than 98% of the incoming wave energy can be dissipated with 20% porosities of the slotted screens and 10% porosity of the porous breakwater.
- The maximum wave-induced forces are observed when $\frac{w}{l} \approx \frac{n}{2}$, $n = 1,2,3, ...$, and the minimum wave-induced forces acting on the porous breakwater are observed in the intermediate points $\frac{(n-1)}{2} < \frac{w}{L} < \frac{n}{2}$, $n = 1,2,3, ...$ for different porosities of the slotted screen.
- A slotted screen of submergence length $\frac{d_1}{h} = \frac{1}{3}$ dissipates the maximum amount of incident wave energy with 10% porosity of the slotted screen.
- The optimum wave-induced horizontal and vertical forces on the floating breakwater depend on the submergence draft of the slotted screen. However, the positions of optima do not depend on the submergence draft of the surface-piercing slotted screen.

The present study can be extended to deal with the seabed undulations as well as irregular water waves. Further, the current solution method can be extended to various structures having complex geometries that arise in ocean engineering.

ACKNOWLEDGEMENTS

SK acknowledges the financial support received through the DST Project: DST/INSPIRE/04/2017/002460 to pursue this research work. Further, supports were received from BITS – Pilani, Hyderabad Campus through BITS/GAU/RIG/2019/H0631 and BITS/GAU/ACRG/2019/H0631(Additional Competitive Research Grant) projects. KPR acknowledges BITS – Pilani, Hyderabad Campus, for providing the financial support to pursue this research work.

REFERENCES

[1]   Wang, C.M., Tay, Z.Y., Takagi, K. & Utsunomiya, T., Literature review of methods for mitigating hydroelastic response of VLFS under wave action. *Applied Mechanics Reviews*, **63**(3), 2010.

[2]   Hong, D.C., Hong, S.Y. & Hong, S.W., Numerical study of the motions and drift force of a floating OWC device. *Ocean Engineering*, **31**(2), pp. 139–164, 2004.

[3]   Tay, Z.Y., Wang, C.M. & Utsunomiya, T., Hydroelastic responses and interactions of floating fuel storage modules placed side-by-side with floating breakwaters. *Marine Structures*, **22**(3), pp. 633–658, 2009.

[4]   Nguyen, H.P., Dai, J., Wang, C.M., Ang, K.K. & Luong, V.H., Reducing hydroelastic responses of pontoon-type VLFS using vertical elastic mooring lines. *Marine Structures*, **59**, pp. 251–270, 2018.

[5]   Ohta, K., Effect of attachment of a horizontal/vertical plate on the wave response of a VLFS. *Proceedings of the 3rd International Workshop on Very Large Floating Structures, VLFS'99*, vol. 1, ed. R. Cengiz Ertekin, Honolulu, Hawaii, USA, Proceedings Paper P1999-5, 1999.

[6]   Watanabe, E., Utsunomiya, T., Kuramoto, M., Ohta, H., Torii, T. & Hayashi, N., Wave response analysis of VLFS with an attached submerged plate. *International Journal of Offshore and Polar Engineering*, **13**(3), 2003.

[7]   Cheng, Y., Ji, C., Zhai, G. & Oleg, G., Dual inclined perforated anti-motion plates for mitigating hydroelastic response of a VLFS under wave action. *Ocean Engineering*, **121**, pp. 572–591, 2016.

[8]   Wang, C.M. et al., Minimizing differential deflection in a pontoon-type, very large floating structure via gill cells. *Marine Structures*, **19**(1), pp. 70–82, 2006.

[9]   Wang, C.M., Riyansyah, M. & Choo, Y.S., Reducing hydroelastic response of interconnected floating beams using semi-rigid connections. *International Conference on Offshore Mechanics and Arctic Engineering*, **43444**, pp. 1419–1425, 2009.

[10]   Gao, R.P., Tay, Z.Y., Wang, C.M. & Koh, C.G., Hydroelastic response of very large floating structure with a flexible line connection. *Ocean Engineering*, **38**(17–18), pp. 1957–1966, 2011.

[11]   Hong, D.C., Hong, S.Y. & Hong, S.W., Reduction of hydroelastic responses of a very-long floating structure by a floating oscillating-water-column breakwater system. *Ocean Engineering*, **33**(5–6), pp. 610–634, 2006.

[12]   Huang, Z., Li, Y. & Liu, Y., Hydraulic performance and wave loadings of perforated/slotted coastal structures: A review. *Ocean Engineering*, **38**(10), pp. 1031–1053, 2011.

[13]   Vijay, K.G., Sahoo, T. & Datta, R., Wave-induced responses of a floating structure near a wall in the presence of permeable plates. *Coastal Engineering Journal*, **62**(1), pp. 35–52, 2020.

[14]   Sun, W.Y., Qu, K., Kraatz, S., Deng, B. & Jiang, C.B., Numerical investigation on performance of submerged porous breakwater to mitigate hydrodynamic loads of coastal bridge deck under solitary wave. *Ocean Engineering*, **213**, p. 107660, 2020.

[15]   Qiao, W., Wang, K.H., Duan, W. & Sun, Y., Analytical model of wave loads and motion responses for a floating breakwater system with attached dual porous side walls. *Mathematical Problems in Engineering*, 2018.

[16]   Yip, T.L. & Chwang, A.T., Perforated wall breakwater with internal horizontal plate. *Journal of Engineering Mechanics*, **126**(5), pp. 533–538, 2000.

[17]   Liu, Y. & Li, H.J., Iterative multi-domain BEM solution for water wave reflection by perforated caisson breakwaters. *Engineering Analysis with Boundary Elements*, **77**, pp. 70–80, 2017.

[18]   Sollitt, C.K. & Cross, R.H., Wave transmission through permeable breakwaters. *Coastal Engineering*, **1972**, pp. 1827–1846, 1973.

[19]   Zhu, S. & Chwang, A.T., Investigations on the reflection behaviour of a slotted seawall. *Coastal Engineering*, **43**(2), pp. 93–104, 2001.

[20]   Koley, S., Panduranga, K., Almashan, N., Neelamani, S. & Al-Ragum, A., Numerical and experimental modeling of water wave interaction with rubble mound offshore porous breakwaters. *Ocean Engineering*, **218**, p. 108218, 2020.

[21]   Li, A.J., Li, H.J. & Liu, Y., Analytical study of oblique wave scattering by a submerged pile-rock breakwater. *Proceedings of the Institution of Mechanical Engineers, Part M: Journal of Engineering for the Maritime Environment*, **233**(1), pp. 41–54, 2019.

[22]   Koley, S. & Panduranga, K., Energy balance relations for flow through thick porous structures. *International Journal of Computational Methods and Experimental Measurements*, **9**(1), pp. 28–37, 2021.

[23]   Hu, J., Zhao, Y. & Liu, P.L.F., A model for obliquely incident wave interacting with a multi-layered object. *Applied Ocean Research*, **87**, pp. 211–222, 2019.

APPLICATION OF A FEM–BEM COUPLING METHOD IN STEADY-STATE HEAT TRANSFER PROBLEM

QI HE[1], YANPENG GONG[1,2] & FEI QIN[1,2]
[1]Institute of Electronics Packaging Technology and Reliability, Faculty of Materials and Manufacturing, Beijing University of Technology, China
[2]Beijing Key Laboratory of Advanced Manufacturing Technology, China

ABSTRACT

In this paper, an FEM–BEM coupling method is presented to study the steady-state heat transfer problem. In the analysis of some complex structures, the FEM–BEM coupling method has many advantages. These advantages are: (i) The degrees of freedom can be highly reduced using the boundary elements; (ii) The improved accuracy of solution over classical FEM; (iii) The use of powerful pre-/post-processing module of the finite element software (ABAQUS); (iv) The improved efficiency of solving multiscale structures. In this work, the user-defined element (UEL) subroutine in ABAQUS is used to realize the coupling of ABAQUS and BEM. The model is divided into several parts and different methods will be used over different part. In the implementation of the coupling scheme, BEM part can be assembled into the ABAQUS as a super-element. And the coupling stiffness matrix which is consisted of the true stiffness matrix of BEM and the stiffness matrix of FEM can be obtained. Then the resulted system equations can be solved by the solver of ABAQUS. Results obtained by this coupling method have an excellent agreement compared with the analytical or reference solutions.
Keywords: FEM–BEM coupling, ABAQUS, UEL, steady-state heat transfer.

1 INTRODUCTION

With the development of science and technology, semiconductor structures are widely used in many areas. Since semiconductor structures are sensitive to the change of temperature, thermal analysis of these structures has been widely studied [1]. In recent years, numerical analysis of semiconductor structures is becoming a hot topic caused by the characteristic of its simplicity, effectiveness and lower cost. Furthermore, the results obtained by numerical techniques can provide some important guidances for the design of semiconductor structures. However, for the numerical analysis of semiconductor structures, large number of multiscale structures usually exist in a numerical model, which results in a big challenge for engineers and researchers [2].

Up to now, many numerical schemes have been used to study multiscale structures including finite element method (FEM), boundary element method (BEM), etc. In the thermal analysis of the electronic devices with multiscale structures by finite element method (FEM), large number of elements is needed to improve the accuracy of the results, which increases the computing time sharply [3]. To solve some practical engineering problems by FEM, many commercial tools are developed such as ABAQUS, ANSYS etc. All these tools can provide powerful pre-/post-processor, which has attracted many engineers. BEM is also a useful numerical technique for the solution of partial differential equations, offering an alternative to the FEM for a range of engineering simulations [4]. The main advantages of the BEM derive from the fact that its approximations (and mesh) only occur on the boundary, the dimension of the numerical model therefore being one less than that of the physical problem being modelled, and from the high accuracy of its solutions on comparatively coarse meshes [5]–[9]. These advantages suggest the BEM can be effectively applied to the analysis of multiscale structures. Actually, the coupling of FEM and BEM is also a powerful method for the thermal analysis of the semiconductor structures. In 1977, Zienkiewicz et al. proposed the coupling of

WIT Transactions on Engineering Sciences, Vol 131, © 2021 WIT Press
www.witpress.com, ISSN 1743-3533 (on-line)
doi:10.2495/BE440101

FEM and BEM to benefit from the combination of two methods [10]. Then, different coupling methods are presented to solve all kinds of problems. In [11], Estorff and Firuziaan apply the coupling BEM/FEM for nonlinear soil/structure interaction. Elleithy et al. proposed the iterative coupling in the elastic statics and elastic plasticity problems [12], [13]. Godinho and Soares also used the coupling method to solve the problem in soil-structure elastic dynamic interaction [14]. In [15], Liu and Dong presented an automatic implementation procedure for the coupling of the ABAQUS with a self-written linear elastic BE code for dynamic elastoplastic problems. There are also many BEM–FEM coupling procedures, applied in nonlinear, dynamic, complex interface and fluid/structure problems, that have been reported in recent years [16]– [19].

In this paper, based on the commercial finite element analysis (FEA) software (ABAQUS), we propose a coupling method of FEM–BEM and apply the scheme to a heat transfer problem. In the implementation of the coupling, the BE region is defined as a large finite element and its stiffness is computed and assembled into the global stiffness matrix of FEM. The stiffness of the model from BE region is computed by the BEM code. To realize the integration of BEM and FEM, the BEM code is put into user-defined element (UEL) subroutine provided by ABAQUS. A classical heat transfer problem is presented to demonstrate the correctness of this method.

2 THEORETICAL ANALYSIS

Many physical problems, such as electromagnetic problems, heat conduction, seepage and acoustic problems, can be described by classical Laplace equations or Poisson equations. For a 2D problem in $\Omega \in \mathbb{R}^2$ with closed boundary Γ, the Laplace equations can be expressed as follows:

$$\nabla^2 u(\mathbf{x}) = \frac{\partial^2 u}{\partial x_1^2} + \frac{\partial^2 u}{\partial x_2^2} = 0, \tag{1}$$

where ∇^2 is the Laplace operator. $u(x)$ is the potential function or temperature at point $\mathbf{x}(x_1, x_2) \in \Omega$.

Eqn. (1) can be solved subject to a set of boundary conditions taken from the following:

$$u(\mathbf{x}) = \bar{u}(\mathbf{x}) \quad \text{on} \quad \mathbf{x} \in \Gamma_1, \tag{2}$$

$$t(\mathbf{x}) = k \frac{\partial u}{\partial \mathbf{n}} = \bar{t}(\mathbf{x}) \quad \text{on} \quad \mathbf{x} \in \Gamma_2, \tag{3}$$

where k is the thermal conductivity; $t(x)$ is the heat flux. $\mathbf{n}$ is the outward pointing normal. The quantities $\bar{u}$ and $\bar{t}$ are known temperature and heat flux, respectively. $\Gamma = \Gamma_1 \cup \Gamma_2$ and $\Gamma_1 \cap \Gamma_2 = \phi$.

2.1 FE formulations for the steady-state heat transfer problem

For the finite element solution of the heat transfer problem, we can obtain the following functional expressions based the conclusion in [20]:

$$\Pi(u) = \int_\Omega \left[\frac{1}{2} k \left(\frac{\partial u}{\partial x_1} \right)^2 + \frac{1}{2} k \left(\frac{\partial u}{\partial x_2} \right)^2 \right] d\Omega - \int_{\Gamma_2} tu d\Gamma. \tag{4}$$

In the implementation of FEM, the considered domain Ω is discretized into surface or body elements. The temperature inside a surface element e can be obtained by following interpolation equation:

$$u \approx \tilde{u} = \sum_{i=1}^{n_e} N_i\,(x_1, x_2) u_i, \tag{5}$$

where n_e is the number of nodes for one element. $N_i(x_1, x_2)$ is the interpolation function. u_i is the local temperature associated with the node with index i.

Substituting eqn (5) into the discrete functional formulation, invoking $\delta\Pi(u) = 0$, we can obtain the following finite element equation for steady heat conduction problems

$$\mathbf{Ku} = \mathbf{R}, \tag{6}$$

where $\mathbf{K}$ is the stiffness matrix for heat transfer problems. Here, $\mathbf{K}$ is a symmetric matrix. Array $\mathbf{u} = [u_1\ u_2\ \cdots\ u_n]^T$ contains the nodal temperature. Vector $\mathbf{R}$ contains the temperature load. The elements K_{ij} and R_i of Matrix $\mathbf{K}$ and $\mathbf{R}$ can be expressed as:

$$K_{ij} = \sum_e K_{ij}^e, \tag{7}$$

$$R_i = \sum_e R_{q_i}^e, \tag{8}$$

where K_{ij}^e and $R_{q_i}^e$ can be expressed as

$$K_{ij}^e = \int_{\Omega^e} k\Big(\frac{\partial N_i}{\partial x_1}\frac{\partial N_j}{\partial x_1} + \frac{\partial N_i}{\partial x_2}\frac{\partial N_j}{\partial x_2}\Big)d\Omega, \tag{9}$$

$$R_{q_i}^e = \int_{\Gamma_2^e} N_i t\, d\Gamma, \tag{10}$$

2.2 BE formulations for the heat transfer problems

For a 2D heat transfer problem, the corresponding boundary integral equation can be written as [21]

$$cu(p) = \int_\Gamma u^*(Q, p) t(Q) d\Gamma(Q) - \int_\Gamma t^*(Q, p) u(Q) d\Gamma(Q), \tag{11}$$

where p represents the source point; Q represents the field point; c is a known value that depends on the geometric shape around the source point p. $u^*(Q, p)$ and $t^*(Q, p)$ denote the temperature and flux fundamental solution kernels, which are defined by

$$u^*(Q, p) = -\frac{1}{2\pi} \ln r, \tag{12}$$

$$t^*(Q, p) = \frac{\partial u^*}{\partial r}\frac{\partial r}{\partial \mathbf{n}}. \tag{13}$$

For the implementation of BEM, the boundary Γ will be divided into n elements Γ_{elem} (elem $= 1, 2, \ldots, n$). Then, the boundary integral eqn (11) can be written in a discretized form

$$c(p_i)u(p_i) + \sum_{\text{elem}=1}^{n} \int_{\Gamma_{\text{elem}}} t^*\,(Q, p_i)\, u^Q d\Gamma\,(Q) = \sum_{\text{elem}=1}^{n} \int_{\Gamma_{\text{elem}}} u^*\,(Q, p_i)\, t^Q d\Gamma\,(Q), \tag{14}$$

where

$$u^Q = \sum_{j=1}^{n_e} N_j u^j, \tag{15}$$

and

$$t^Q = \sum_{j=1}^{n_e} N_j t^j. \tag{16}$$

Γ_{elem} is the boundary of element with index 'elem'; p_i is the source point with index i.

By considering eqn (14) at a sufficient number of source points, we can get the system of equation

$$\mathbf{Hu} = \mathbf{Gt}, \tag{17}$$

where $\mathbf{u}$ and $\mathbf{t}$ represent the vectors contained temperature and flow at boundary nodes. Matrix $\mathbf{H}$ is a square matrix containing a combination of the integrals of the t^* kernel and coefficient c. $\mathbf{G}$ is a rectangular matrix of u^* kernel integrals.

3 COUPLING OF BEM AND FEM

Considering the coupling of structure domain Ω^F and Ω^B ($\Omega = \Omega^F \cup \Omega^B$, $\Omega^F \cap \Omega^B = \phi$). In domain Ω^F, the finite element method will be used. And boundary element method will be used in domain Ω^B.

3.1 Formulations of finite element method

According to the nodal locations, eqn (7) can be written as

$$\begin{bmatrix} \mathbf{K}_{oo} & \mathbf{K}_{oi} \\ \mathbf{K}_{io} & \mathbf{K}_{ii} \end{bmatrix} \begin{Bmatrix} \mathbf{t}_{F_o} \\ \mathbf{u}_{F_i} \end{Bmatrix} = \begin{Bmatrix} \mathbf{R}_{F_o} \\ \mathbf{R}_{F_i} \end{Bmatrix}, \tag{18}$$

where $\mathbf{K}_{sub}$, $\mathbf{t}_{sub}$, $\mathbf{u}_{sub}$, and $\mathbf{R}_{sub}$ are the newly constructed matrices and vectors. F_i represents the quantities related to the interface in the FEM domain. And F_o indicates the quantities related to the non-interface part in the FEM domain.

3.2 Formulations of boundary element method

Similarly, based on the nodal locations, eqn (17) can be written as

$$\begin{bmatrix} \mathbf{H}_{ii} & \mathbf{H}_{io} \\ \mathbf{H}_{oi} & \mathbf{H}_{oo} \end{bmatrix} \begin{Bmatrix} \mathbf{u}_{B_i} \\ \mathbf{u}_{Bo} \end{Bmatrix} = \begin{bmatrix} \mathbf{G}_{ii} & \mathbf{G}_{io} \\ \mathbf{G}_{oi} & \mathbf{G}_{oo} \end{bmatrix} \begin{Bmatrix} \mathbf{t}_{B_i} \\ \mathbf{t}_{Bo} \end{Bmatrix}, \tag{19}$$

where $\mathbf{H}_{sub}$, $\mathbf{G}_{sub}$, $\mathbf{u}_{sub}$ and $\mathbf{t}_{sub}$ are the newly constructed matrices and vectors. And the subscript B_i represents the quantities related to the interface in the BEM domain. B_o indicates the quantities related to the non-interface part in the BEM domain.

Along the interface between the domains Ω^F and Ω^B, the continuity condition requires that temperature calculated for the Ω^F must equal the temperature calculated for the Ω^B. And a similar relationship remains for the equilibrium condition along the interface, except that a negative sign must be given to account for the opposite directions of the outward boundary normal in the two domains. To use these relationships, eqn (19) can be transformed to the following form by setting $\mathbf{H}_{io} = 0$

$$\begin{bmatrix} \mathbf{A}_{ii} & 0 \\ \mathbf{A}_{oi} & \mathbf{A}_{oo} \end{bmatrix} \begin{Bmatrix} \mathbf{u}_{B_i} \\ \mathbf{u}_{Bo} \end{Bmatrix} = \begin{bmatrix} \mathbf{B}_{ii} & \mathbf{B}_{io} \\ \mathbf{B}_{oi} & \mathbf{B}_{oo} \end{bmatrix} \begin{Bmatrix} \mathbf{t}_{B_i} \\ \mathbf{t}_{Bo} \end{Bmatrix}. \tag{20}$$

From eqn (20), we can obtain the following equation

$$\mathbf{A}_{ii}\mathbf{u}_{B_i} = \mathbf{B}_{ii}\mathbf{t}_{B_i} + \mathbf{B}_{io}\mathbf{t}_{Bo}. \tag{21}$$

Then, eqn (21) can be expressed as

$$\mathbf{t}_{B_i} = \mathbf{B}_{ii}^{-1}\mathbf{A}_{ii}\mathbf{u}_{B_i} - \mathbf{B}_{ii}^{-1}\mathbf{B}_{io}\mathbf{t}_{Bo}. \tag{22}$$

The heat flux $\mathbf{t}_{B_i}$ in eqn (22), should be converted into the equivalent nodal flux $\mathbf{R}_{B_i}$ which is used in finite element method. And $\mathbf{R}_{B_i}$ can be expressed in the form

$$\mathbf{R}_{B_i} = \mathbf{M}_B\mathbf{t}_{B_i} = \mathbf{K}_{B_i}\mathbf{u}_{B_i} - \bar{\mathbf{R}}_{B_i}, \tag{23}$$

where $\mathbf{M}_B$ is a transformation matrix and

$$\bar{\mathbf{R}}_{B_i} = \mathbf{M}_B\mathbf{B}_{ii}^{-1}\mathbf{B}_{io}\mathbf{t}_{Bo}. \tag{24}$$

According to the continuity condition of interface, the coupling equation can be finally expressed as

$$\begin{bmatrix} \mathbf{K}_{oo} & \mathbf{K}_{oi} \\ \mathbf{K}_{io} & \mathbf{K}_{ii} + \mathbf{K}_{B_i} \end{bmatrix} \left\{ \begin{array}{c} \mathbf{u}_{F_o} \\ \mathbf{u}_{F_i} \end{array} \right\} = \left\{ \begin{array}{c} \mathbf{R}_{F_o} \\ \bar{\mathbf{R}}_{B_i} \end{array} \right\}. \tag{25}$$

It should be noted that $\mathbf{K}_{B_i}$ obtained from BEM domain is asymmetric, which destroys the sparsity of the FE stiffness matrix. Therefore, we will replace $\mathbf{K}_{B_i}$ by an enhanced symmetric matrix $\mathbf{K}_l$. And the coupling equation in eqn (25) is converted as follows:

$$\begin{bmatrix} \mathbf{K}_{oo} & \mathbf{K}_{oi} \\ \mathbf{K}_{io} & \mathbf{K}_{ii} + \mathbf{K}_l \end{bmatrix} \left\{ \begin{array}{c} \mathbf{u}_{F_o} \\ \mathbf{u}_{F_i} \end{array} \right\} = \left\{ \begin{array}{c} \mathbf{R}_{F_o} \\ \bar{\mathbf{R}}_{B_i} \end{array} \right\} + \left\{ \begin{array}{c} 0 \\ -\mathbf{K}_{B_i} + \mathbf{K}_l \end{array} \right\} \{\mathbf{u}_{F_i}\}, \tag{26}$$

where

$$\mathbf{K}_l = \frac{1}{2}\left(\mathbf{K}_{B_i}^T + \mathbf{K}_{B_i}\right). \tag{27}$$

4 IMPLEMENTATION OF THE COUPLING WITH ABAQUS

ABAQUS is a powerful engineering finite element software, including various types of element libraries and material model libraries. It is becoming an important tool for engineers to solve complex nonlinear problems. In this section, we will introduce the scheme that combines a self-written BEM-code and the commercial finite element software ABAQUS. The UEL (User Element Subroutine) [22] provided by ABAQUS will be used to realize the coupling of FEM and BEM. The implementation process is given as follows:

- As shown in Fig. 1(a), the considered domain is divided into two sub-regions, i.e. FE region and BE region. Then, the FE region needs to be divided into quadrilateral or triangular finite elements which can be easily solved by pre-process tools of ABAQUS. While, for the BE region, only boundary element is used (see Fig. 1(b)).
- In the coupling scheme, one single model including FE and BE parts (shown in Fig. 2(a)) should be built by the ABAQUS platform. Here the BE part is defined as a one-dimensional simplified FE part. It should be noted that the number of nodes at the interface between FE region and BE region must be equal. And the corresponding nodes will be connected by the command 'tie', as shown in Fig. 2(b). Then the model can be analyzed as a pure FE model.

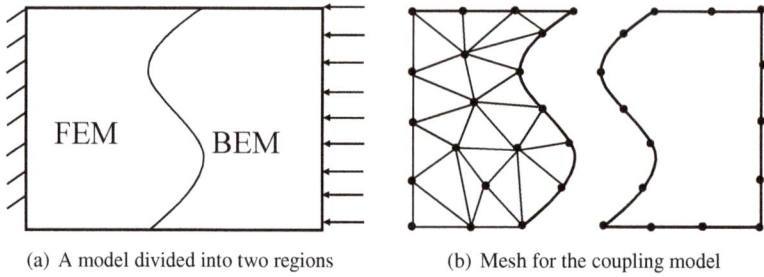

(a) A model divided into two regions   (b) Mesh for the coupling model

Figure 1: The coupling model.

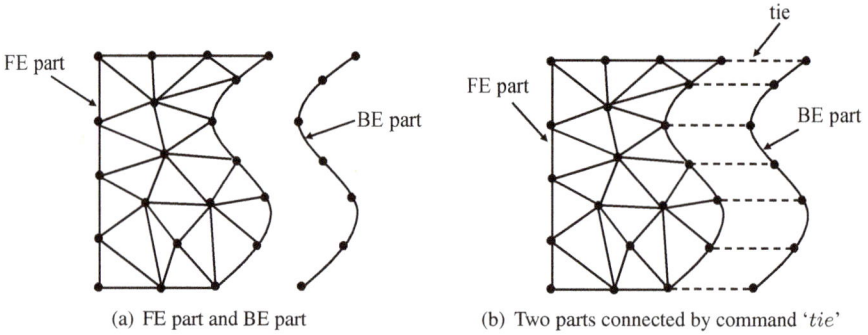

(a) FE part and BE part   (b) Two parts connected by command '*tie*'

Figure 2: Model built by ABAQUS.

5 NUMERICAL EXAMPLE

To demonstrate the accuracy and effectiveness of the new coupling approach, we examine a simple example about heat transfer problem. To carry out the accuracy and convergence analysis, a relative error defined in eqn (28)

$$\text{Relative error (RE)} = \frac{|f_{\text{num}} - f_{\text{ref}}|}{|f_{\text{ref}}|}, \tag{28}$$

where f_{num} and f_{ref} represent the numerical solution and reference solution, respectively.

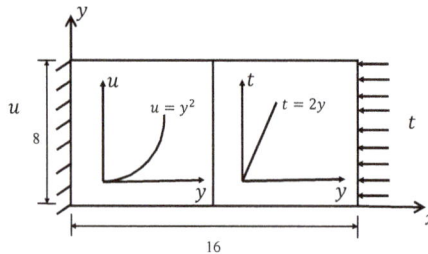

Figure 3: Considered rectangular plate.

As shown in Fig. 3, consider a 8×16 rectangular plate. The Dirichlet boundary condition is exerted on the left side of the plate and given by the $u = y^2$. The Neumann BC is exerted on the right side of the right region. Its given by the $t = 2y$. All the upper and lower boundaries are insulated, i.e. $\Delta u \cdot \mathbf{n} = 0$. In the computation by the current coupling method, the plate in Fig. 3 is divided into two parts. And the left domain will be analyzed by FEM and the BEM will be used on the right part.

The mesh used in the coupling method is given in Fig. 4(a), from which we can see that the left region is divided into quadrilateral elements. And only the boundary is discretized in the right region. Fig. 4(b) shows the one-dimensional simplified FE part. And we can see from Fig. 4(b) that the number of nodes on the left side equals the number of nodes for the right curve.

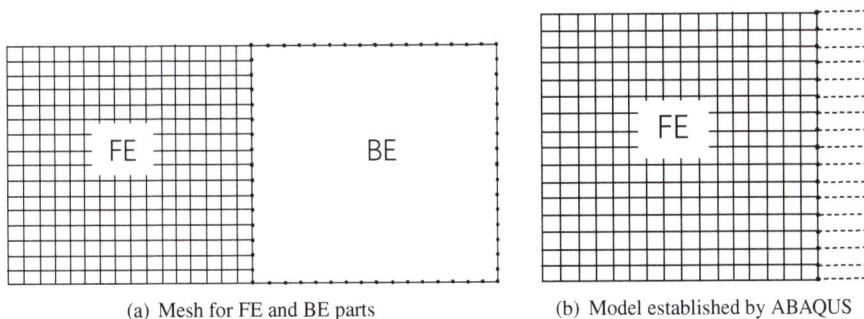

(a) Mesh for FE and BE parts   (b) Model established by ABAQUS

Figure 4: Mesh for the considered model.

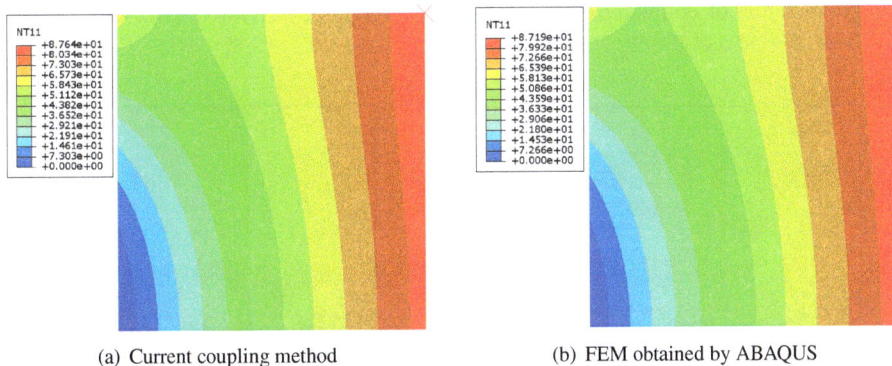

(a) Current coupling method   (b) FEM obtained by ABAQUS

Figure 5: Temperature distribution for different methods.

Here, a finite element model is constructed with ABAQUS to offer a reference solution. The element DC2D4 is used in this numerical example. The mesh consists of 1024 elements and 1089 nodes. The temperature distribution of the left region is given in Fig. 5(a) and 5(b) for the two methods. One can find that the results obtained by the current method are in good agreement with the FEM solutions.

To study the accuracy of the present method, some points are selected along the bottom boundary (with parametric equations $(x, 0)$, $0 \leq x \leq 8$) and the interface between the FEM part and BEM part. The temperature along different curves for different ndof are given in

Fig. 6(a) and 6(b). Here, the number of degree of freedom (ndof) is the sum of ndof from FE part and BE part.

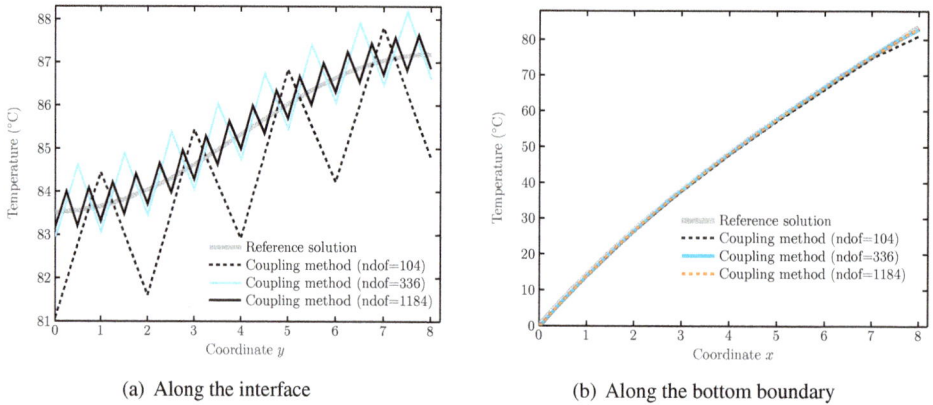

(a) Along the interface

(b) Along the bottom boundary

Figure 6: The temperature calculated along different curves for different ndofs.

To study the convergence rate of the present method, a problem for which we can find the analytical solution is studied. The Dirichlet boundary condition on the left is changed as the fixed value 0. And the Neumann boundary condition is changed to 1. The exact temperatures for any point can be given as $u = x$. Fig. 7(a) and 7(b) show the relative errors of temperature as the ndof increases from 336 to 1184. From Fig. 7(a) and 7(b), we draw the same conclusions as from Fig. 6(a) and 6(b), in that the relative errors of interfacial points are larger than that of boundary points. And the convergence of relative errors can be clearly seen for the two sets of points.

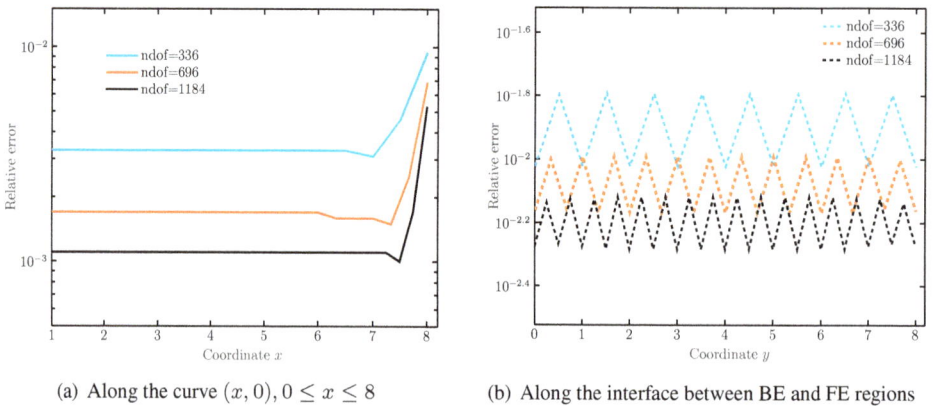

(a) Along the curve $(x, 0), 0 \leq x \leq 8$

(b) Along the interface between BE and FE regions

Figure 7: Relative errors of temperature for different ndof.

6 CONCLUSION

Based on the finite element software ABAQUS, this paper presents a FEM–BEM coupling method and apply it to the steady-state heat transfer problem. In the implementation of coupling method, the UEL provided by ABAQUS is used to combine the self-written BEM-code and ABAQUS. The self-written BEM-code is used to obtain the effective stiffness matrix of the BE region. Then the resulted effective stiffness matrix and stiffness matrix of FEM region will form a coupling stiffness matrix, which will be solved by the solver of ABAQUS. Finally, a simple example about heat transfer problem is calculated by this method.

ACKNOWLEDGEMENTS

The research was supported by the National Natural Science Foundation of China (No. 12002009), the General Program of Science and Technology Development Project of Beijing Municipal Education Commission (No. KM202110005032), Beijing Postdoctoral Research Foundation and General Program of Science and Technology Development Project of BJUT.

REFERENCES

[1] Bailey, C., Thermal management technologies for electronic packaging: Current capabilities and future challenges for modelling tools. *2008 10th Electronics Packaging Technology Conference*, IEEE, pp. 527–532, 2008.

[2] Ladeveze, P., Multiscale modelling and computational strategies for composites. *International Journal for Numerical Methods in Engineering*, **60**(1), pp. 233–253, 2004.

[3] Blackburn, D.L., Temperature measurements of semiconductor devices – A review. *Twentieth Annual IEEE Semiconductor Thermal Measurement and Management Symposium (IEEE Cat. No. 04CH37545)*, IEEE, pp. 70–80, 2004.

[4] Nardini, D. & Brebbia, C., A new approach to free vibration analysis using boundary elements. *Applied Mathematical Modelling*, **7**(3), pp. 157–162, 1983.

[5] Gu, Y., Chen, W. & Zhang, C., Stress analysis for thin multilayered coating systems using a sinh transformed boundary element method. *International Journal of Solids and Structures*, **50**(20–21), pp. 3460–3471, 2013.

[6] Gu, Y., He, X., Chen, W. & Zhang, C., Analysis of three-dimensional anisotropic heat conduction problems on thin domains using an advanced boundary element method. *Computers & Mathematics with Applications*, **75**(1), pp. 33–44, 2018.

[7] Baynham, J., Adey, R., Murugaian, V. & Williams, D., Simulating electro-coating of automotive body parts using BEM. *WIT Transactions on Engineering Sciences*, **54**, 2007.

[8] Zhou, H.L., Han, H.S., Cheng, C.Z. & Niu, Z.R., Inverse identification of heat boundary conditions for 2-D anisotropic coating structures. *Applied Mechanics and Materials, Trans Tech Publications*, **130**, pp. 1825–1828, 2012.

[9] Zhang, Y.M., Gu, Y. & Chen, J.T., Stress analysis for multilayered coating systems using semi-analytical BEM with geometric non-linearities. *Computational Mechanics*, **47**(5), pp. 493–504, 2011.

[10] Zienkiewicz, O., Kelly, D. & Bettess, P., The coupling of the finite element method and boundary solution procedures. *International Journal for Numerical Methods in Engineering*, **11**(2), pp. 355–375, 1977.

[11] Von Estorff, O. & Firuziaan, M., Coupled BEM/FEM approach for nonlinear soil/structure interaction. *Engineering Analysis with Boundary Elements*, **24**(10), pp. 715–725, 2000.

[12] Elleithy, W.M., Al-Gahtani, H.J. & El-Gebeily, M., Iterative coupling of BE and FE methods in elastostatics. *Engineering Analysis with Boundary Elements*, **25**(8), pp. 685–695, 2001.

[13] Elleithy, W. & Grzhibovskis, R., An adaptive domain decomposition coupled finite element–boundary element method for solving problems in elasto-plasticity. *International Journal for Numerical Methods in Engineering*, **79**(8), pp. 1019–1040, 2009.

[14] Godinho, L. & Soares Jr., D., Numerical simulation of soil-structure elastodynamic interaction using iterative-adaptive BEM-FEM coupled strategies. *Engineering Analysis with Boundary Elements*, **82**, pp. 141–161, 2017.

[15] Liu, Z. & Dong, C., Automatic coupling of ABAQUS and a boundary element code for dynamic elastoplastic problems. *Engineering Analysis with Boundary Elements*, **65**, pp. 147–158, 2016.

[16] Soares Jr., D., FEM-BEM iterative coupling procedures to analyze interacting wave propagation models: Fluid-fluid, solid-solid and fluid-solid analyses. *Coupled systems Mechanics*, **1**(1), p. 19, 2012.

[17] Pavlatos, G. & Beskos, D., Dynamic elastoplastic analysis by BEM/FEM. *Engineering Analysis with Boundary Elements*, **14**(1), pp. 51–63, 1994.

[18] Chien, C.C. & Wu, T.Y., A particular integral BEM/time-discontinuous FEM methodology for solving 2-D elastodynamic problems. *International Journal of Solids and Structures*, **38**(2), pp. 289–306, 2001.

[19] Boumaiza, D. & Aour, B., On the efficiency of the iterative coupling FEM–BEM for solving the elasto-plastic problems. *Engineering Structures*, **72**, pp. 12–25, 2014.

[20] Zienkiewicz, O.C., Taylor, R.L. & Zhu, J.Z., *The Finite Element Method: Its Basis and Fundamentals*, Elsevier, 2005.

[21] Gong, Y., Trevelyan, J., Hattori, G. & Dong, C., Hybrid nearly singular integration for three-dimensional isogeometric boundary element analysis of coatings and other thin structures. *Computer Methods in Applied Mechanics and Engineering*, **367**, pp. 642–673, 2019.

[22] Hibbitt, Karlsson & Sorensen, *ABAQUS/Standard: User's Manual*, vol. 1, Hibbitt, Karlsson & Sorensen, 1997.

ON THE APPLICATION OF HYBRID BEM/FEM FOR INHOMOGENEOUS ELECTROMAGNETIC SCATTERING FROM SMALL MARITIME TARGETS

HRVOJE DODIG[1], DRAGAN POLJAK[2] & MARIO CVETKOVIĆ[2]
[1]Faculty of Maritime Studies, University of Split, Croatia
[2]Faculty of Electrical Engineering, Mechanical Engineering and Naval Architecture, University of Split, Croatia

ABSTRACT

From the standpoint of detectability of small maritime objects by a ship's radar system, the radar cross section (RCS) of a maritime object is a key parameter that determines the detectability of an object. When a vessel navigates in narrow channels and in areas with high sea traffic density, the ability to detect small maritime targets such as buoys and inflatable rubber boats is crucial for maritime safety. However, it is difficult to measure the radar cross section of such maritime objects in environments such as laboratory or in a field test not only due to object size, but also due to the fact that these objects are partially submerged into seawater. In such situations, the most effective approach is to apply numerical methods for electromagnetic scattering from inhomogeneous objects in order to predict the RCS of a maritime object. Though approaches to RCS prediction are diverse, one of the most sophisticated, namely, the hybrid numerical method based on combination of boundary and finite elements is applied in this work. This hybrid numerical method first computes near electromagnetic field on the surface of the scatterer, followed by a particular technique developed by the authors to compute the radar cross section of small maritime targets from these near field values. The accuracy of the approach is discussed and some illustrative computational examples of RCS calculations are given.
Keywords: boundary element method, finite element methods, hybrid boundary element/finite element method, electromagnetic scattering, radar cross section, small maritime targets.

1 INTRODUCTION

In coastal waters near tourist areas where dense maritime traffic is expected, the ability to detect small vessels such as inflatable rubber boats and buoys is of high importance. During night cruises or in bad weather conditions, the only sensor available to pilots and captains that can detect such objects is radar. The detection probability of these small maritime targets is directly proportional to an electrical engineering parameter called radar cross section (RCS). The radar cross section represents the susceptibility of radar target to detection by a radar [1], and it is the parameter that represents the ratio of backscattered power to transmitted power. Consequently, the lower the RCS, the smaller is probability of detection of radar target by a ship's radar.

While RCS does not necessarily depend on target's physical size, the most important factors are target's geometry and material composition. For example, radar-deflectors, sometimes called radar target enhancers (RTE), reflect radar antenna energy in such a way that the target appears larger on radar's display [2]. RTEs are physically small in size and their typical physical dimensions vary from 8 cm to 20 cm. These devices are made of metal, and their geometry is designed in such a way that it enhances the RCS. On the other hand, inflatable rubber boats typically consist of materials such as plastic, wood and rubber which contribute very little to RCS. This impairs the detection probability of rubber boat by ships radar, thus, many of these inflatable rubber boats install radar reflectors to improve the navigational safety and to decrease the chance of collision with larger vessel.

In the design of radar systems with the ability to detect these small maritime objects it is advantageous to know in advance the RCS to be expected. It is very difficult to realistically

WIT Transactions on Engineering Sciences, Vol 131, © 2021 WIT Press
www.witpress.com, ISSN 1743-3533 (on-line)
doi:10.2495/BE440111

measure RCS of rubber boats and buoys in laboratory conditions because of the sheer size of these objects and because these objects are partially submerged into seawater. On the other hand, if field measurements are performed, the difficulties in RCS measurement arise from not being able to control weather conditions and the sea state such as waves and currents. Hence, perhaps the most cost effective and reliable solution to RCS prediction is to numerically calculate RCS of these small maritime objects.

Numerical methods that can be applied in order to predict the RCS of small maritime target generally fall into one of the following categories: ray-tracing methods [3], [4], ray-tracing combined with physical optics methods (PO) or with physical theory of diffraction (PTD) [5], [6], and full wave methods based on hybrid method of moments/finite element method (MoM/FEM) or hybrid boundary element method/finite element method (BEM/FEM) [7]. Because of multiple reflections of rays between facets [8], ray tracing methods are sometimes difficult to apply if the geometry of the target has highly irregular features. Raytracing combined with PO or PTD can produce RCS calculation with acceptable error for homogeneous scatterers, however, these methods can have difficulties when there are material inhomogeneities within the penetration depth of radar wave.

Full wave methods that can be used for RCS prediction are FDTD [9], hybrid MoM/FEM [10], volume-surface integral equation (VSIE) and hybrid BEM/FEM [7]. These methods come with the cost of increased computational burden when compared to raytracing with PO/PTD methods. However, they can properly take into account inhomogeneities of target's material composition. FDTD can sometimes have problems with irregular surfaces because of staircasing. VSIE methods combine surface integral equation with volume integral equation and MoM is typically used to assemble the linear system of equations. Consequently VSIE produces dense matrices for both surface and interior of the computational problem it is impractical for RCS computation of anything but the smallest objects.

Hybrid MoM/FEM method combines surface integral equation with FEM and it uses MoM to create a linear system of equations for the surface of electromagnetic scattering problem. Hence, MoM/FEM generates the linear system of equations which is sparse in the interior of the computational problem (because of FEM) and dense at the bounding surface of the computational problem (because of MoM). In the hybrid MoM/FEM method, typical elements used for surface discretization of an EM scattering problem are Rao–Wilton–Glisson (RWG) elements. This can sometimes pose a difficulty because these elements are not divergenceless which is physically unrealistic; however, that method approaches non-zero divergence condition with increased surface discretization.

Thus, in order to implement a physically correct method for EM scattering computation which can take inhomogeneities of materials into account, in this paper we use the hybrid BEM/FEM method with edge elements. Edge elements are used in hybrid BEM/FEM for electromagnetic scattering to model both interior of the problem (FEM part) and the surface of the problem (BEM part) which is an advantage in itself because the same type of elements is used to model interior and the surface of EM scattering problem. Furthermore, this type of elements bear two advantages: (i) edge elements are divergenceless and (ii) they automatically preserve tangential continuity of electromagnetic fields. Tangential continuity of electromagnetic fields is a physically required condition on the surfaces that are interface between materials with different electrical properties. Additional advantage of edge elements is that all the surface integrals in EM scattering computation with hybrid BEM/FEM that come from BEM can be converted to contour integrals [11]. Controlling the accuracy of the calculation of BEM integrals is significantly simpler when these integrals are computed as contour integrals instead of surface integrals. Finally, regarding hybrid BEM/FEM with edge elements, it should be noted that the research on this type of elements used in hybrid

BEM/FEM started a long time ago, however, due to some mathematical difficulties the method did not became main-stream method compared to BEM/FEM. The resolution to some of these difficulties is reported in [12].

Regardless of the applied full wave method, the final result of numerical EM scattering computation is near electromagnetic field. This near electrical field needs to be converted to far field in order to compute the RCS. Standard near-to-far field transformation (NTFFT) is a computationally cumbersome procedure [13]. To circumvent this, the approach based on [14] is applied to compute the far field and RCS. The method is described in Section 3, and it allows rapid computation of RCS directly from edge element coefficients provided by near field computation.

In Section 4, the solid models of buoy and rubber boat, respectively, are given. While the buoy model was taken from technical specification of the buoy, the model of rubber boat is constructed by 3D laser scanning (courtesy of Croatian Navy). Finally, the results of numerical computation of the RCS of the buoy and the rubber boat are presented in Section 5 for both horizontal polarization (HP) and vertical polarization (VP) of EM wave. These results were compared to RCS of the sea itself in an attempt to predict the probability of detection.

2 HYBRID BEM FEM FOR ELECTROMAGNETIC SCATTERING

The first step in computation of RCS of radar target is to compute backscattered electric and magnetic field at the bounding surface ∂V of the computational problem shown in Fig. 1. If the surface ∂V is near the target then the computed electromagnetic field will be near electromagnetic field. In Fig. 1 the incident electromagnetic field denoted $\vec{E}_i$ and $\vec{H}_i$ excites

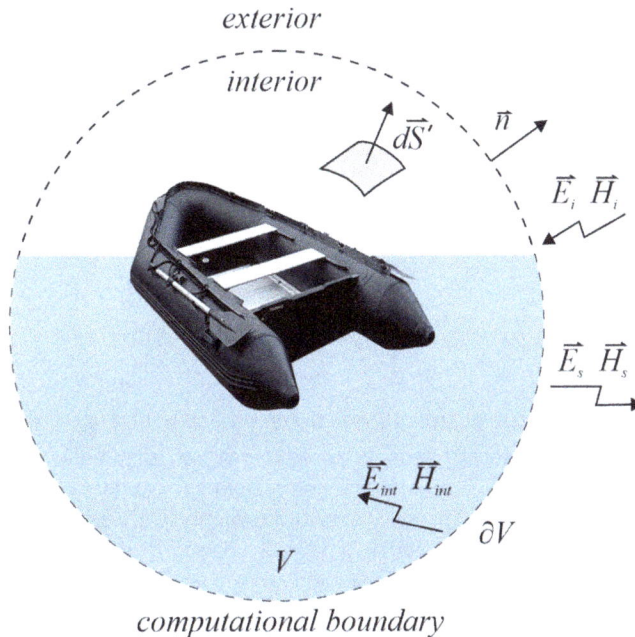

Figure 1:   Outline of EM scattering problem. Volume of computational domain is denoted V and the artificial boundary is denoted ∂V. Fields $\vec{E}_i$ and $\vec{H}_i$ are incident to ∂V while $\vec{E}_S$ and $\vec{H}_S$ are backscattered fields.

the sources on the surface of the scatterer which in turn generate backscattered electromagnetic field denoted $\vec{E}_S$ and $\vec{H}_S$. Fields interior to the computational domain are denoted $\vec{E}_{int}$ and $\vec{H}_{int}$. For the purpose of RCS computation we need to compute backscattered fields $\vec{E}_S$ and $\vec{H}_S$. However, the backscattered fields are affected by field distribution inside the scatterer and because the scatterer is considered to be inhomogeneous electromagnetic properties of materials ($\epsilon_r, \mu_r, \sigma$) change inside computational domain V. Hence, to compute backscattered fields we need the computational method that can take into account this change of electromagnetic properties of materials.

The method that can properly take into account the changing electromagnetic properties of materials inside computational domain is hybrid BEM/FEM which is thoroughly described in references [12], [15], [16]. The starting point for hybrid BEM/FEM is magnetic field surface integral equation for time harmonic electromagnetic fields [17]:

$$
\alpha \vec{H}_{ext} = \vec{H}_i + j\omega\epsilon \oint_{\partial V} d\vec{S}' \times \vec{E}'_{ext}\, G + \oint_{\partial V} \left(d\vec{S}' \times \vec{H}'_{ext}\right) \times \nabla' G
$$
$$
- \frac{j}{\omega\mu} \oint_{\partial V} dS' \nabla_S' \cdot \left(\vec{n}' \times \vec{H}'_{ext}\right) \nabla' G. \tag{1}
$$

In eqn (1), fields $\vec{E}_{ext}$ and $\vec{H}_{ext}$ denote the electric and magnetic fields exterior to the computational domain shown in Fig. 1. Fields interior to the computational domain (denoted $\vec{E}_{int}$ and $\vec{H}_{int}$) are governed by Maxwell equations. Two Maxwell equations that are necessary to implement FEM part of hybrid BEM/FEM method are Faraday's law and Maxwell–Ampere equation which can be written in time-harmonic form [18]:

$$
\nabla' \times \vec{E}_{int} = -j\omega\mu\vec{H}_{int}, \tag{2}
$$

$$
\nabla' \times \vec{H}_{int} = (\sigma + j\omega\epsilon)\vec{E}_{int.} \tag{3}
$$

Dividing eqn (2) by $-j\omega\mu$, and then taking the curl of eqn (2) and combining with eqn (3) yields:

$$
\nabla \times \left(\frac{j}{\omega\mu}\nabla \times \vec{E}_{int}\right) - (\sigma + j\omega\epsilon)\vec{H}_{int} = 0. \tag{4}
$$

Let us now suppose that computational domain shown in Fig. 1 is discretized into nonconformal mesh consisting of tetrahedral elements and that fields $\vec{E}_{int}$ and $\vec{H}_{int}$ are approximated by some type of elements. The nodal elements cannot be used for this type of computation because these would produce spurious (nonphysical) results. However, one can use tangential edge elements [19] which physically correctly model the electromagnetic fields inside each tetrahedron. These type of elements are divergenceless which is physically required condition over each tetrahedron. Furthermore, edge elements preserve tangential continuity of electromagnetic fields across inter-element faces which is again physically required condition. Over each element, electric and magnetic fields can be approximated using edge elements as:

$$\vec{E}_{int} = \sum_{k=1}^{n} \delta_k \vec{w}_k e_k, \tag{5}$$

$$\vec{H}_{int} = \sum_{k=1}^{n} \delta_k \vec{w}_k h_k, \tag{6}$$

where $\vec{w}_k$ is vector approximating function, n is the number of edges on the element, e_k and h_k are unknown coefficients associated with each edge of the element. Vector approximating functions $\vec{w}_k$ are associated with k^{th} edge of the element by the following relation:

$$\vec{w}_k = N_i \nabla N_j - N_j \nabla N_i, \tag{7}$$

where N_i and N_j are usual nodal approximating functions associated with endpoint nodes of kth element edge. With edge elements every global edge in the nonconformal tetrahedral mesh bust be given a global direction. This direction might differ from local direction of each edge and coefficient δ_k is either $+1$ or -1 depending on whether local edge direction coincides with global edge direction.

To apply FEM the Galerkin method is used. Galerkin method requires that the volume integral of vector dot product of $\delta_k \vec{w}_k$ and eqn (4) vanishes over the volume V:

$$\int_V \delta_k \vec{w}_k \cdot \left[\nabla \times \left(\frac{j}{\omega \mu} \nabla \times \vec{E}_{int} \right) - (\sigma + j\omega\epsilon) \vec{H}_{int} \right] dV = 0. \tag{8}$$

Using standard vector identity $\nabla \cdot \vec{P} \times \vec{Q} = \vec{Q} \cdot \nabla \times \vec{P} - \vec{P} \cdot \nabla \times \vec{Q}$ and divergence theorem the weak formulation is obtained:

$$\int_V \frac{j}{\omega \mu} \nabla \times \delta_k \vec{w}_k \cdot \nabla \times \vec{E} \, dV + \int_V (\sigma + j\omega\epsilon) \delta_k \vec{w}_k \cdot \vec{E} \, dV = \oint_{\partial V} d\vec{S} \cdot \delta_k \vec{w}_k \times \vec{H}. \tag{9}$$

Note that notation $\vec{E}_{int}$ and $\vec{H}_{int}$ is dropped because of simplicity and that symbols $\vec{E}$ and $\vec{H}$ now denote the fields interior to volume V.

To convert exterior fields $\vec{E}_{ext}$ and $\vec{H}_{ext}$ in eqn (1) to interior fields $\vec{E}$ and $\vec{H}$ we make use of physical requirement that $\vec{n} \times \vec{H}_{ext} = \vec{n} \times \vec{H}, \vec{n} \times \vec{E}_{ext} = \vec{n} \times \vec{E}$. Furthermore, in order to treat boundary integral eqn (1) with FEM we take the cross product of whole eqn (1) with unit surface normal $\vec{n}$ (normal to surface ∂V) to obtain:

$$\alpha \vec{n} \times \vec{H} = \vec{n} \times \vec{H}_i + j\omega\epsilon \vec{n} \times \oint_{\partial V} d\vec{S}' \times \vec{E} \, G + \vec{n} \times \oint_{\partial V} \left(d\vec{S}' \times \vec{H} \right) \times \nabla' G$$
$$- \frac{j}{\omega \mu} \vec{n} \times \oint_{\partial V} dS' \nabla_s' \cdot \left(\vec{n}' \times \vec{H} \right) \nabla' G. \tag{10}$$

Finally, to produce a linear system of equations, the eqn (10) is treated with Galerkin FEM by requiring that surface integral of dot product eqn (10) with interpolating functions $\delta_k \vec{w}_k$ vanishes:

$$\int_{\partial V} dS \, \alpha \delta_k \vec{w}_k \cdot \vec{n} \times \vec{H}$$

$$= \oint_{\partial V} dS \delta_k \vec{w}_k \cdot \vec{n} \times \vec{H}_i + j\omega\epsilon \oint_{\partial V} dS \delta_k \vec{w}_k \cdot \oint_{\partial V} d\vec{S}' \times \vec{E}\,G$$

$$+ \oint_{\partial V} dS \delta_k \vec{w}_k \cdot \vec{n} \times \oint_{\partial V} \left(d\vec{S}' \times \vec{H}\right) \times \nabla' G$$

$$- \frac{j}{\omega\mu} \oint_{\partial V} dS \delta_k \vec{w}_k \cdot \vec{n} \times \oint_{\partial V} dS' \nabla_S' \cdot \left(\vec{n}' \times \vec{H}\right) \nabla' G. \tag{11}$$

Using standard FEM procedures to eqns (9) and (11) one can now obtain a linear system of equations written in matrix form as:

$$\begin{bmatrix} H & G & 0 \\ D & M & M \\ 0 & M & M \end{bmatrix} \begin{Bmatrix} h_b \\ e_b \\ e_v \end{Bmatrix} = \begin{Bmatrix} h_I \\ 0 \\ 0 \end{Bmatrix}, \tag{12}$$

where e_I are edge element coefficients computed from incident field $\vec{H}_i$ (using first right hand side term of eqn (11)), coefficients h_b and e_b are edge element coefficients associated with edges on computational boundary ∂V and e_v are edge element coefficients associated with edges that are inside computational domain V (and not on computational boundary ∂V). By solving the system of eqn (12) one can recreate electric field inside each tetrahedron in computational domain V using eqn (5). However, for RCS computation we are only interested in coefficients h_b and e_b because these are associated with edges on computational boundary ∂V. Regarding the testing of this hybrid BEM/FEM computational method, the method was rigorously tested in various physical settings over the course of several years (see [12], [15], [16], [20]).

3 COMPUTATION OF RADAR CROSS SECTION

Once the edge element coefficients h_b and e_b are computed one can directly compute RCS from these coefficients without transforming near electromagnetic fields to far fields by using NTFFT by implementing technique developed by authors of this paper [14]. Radar cross scattering section σ is defined as ratio of backscattered and incident field:

$$\sigma = \lim_{|\vec{r}|\to\infty} 4\pi|\vec{r}|^2 \frac{\left|\vec{E}_S\right|^2}{\left|\vec{E}_i\right|^2} = \lim_{|\vec{r}|\to\infty} 4\pi|\vec{r}|^2 \frac{\vec{E}_S \cdot \vec{E}_S^*}{\vec{E}_I \cdot \vec{E}_I^*}, \tag{13}$$

where $\vec{E}_S^*$ represents the complex conjugate of backscattered vector field $\vec{E}_S$. In Ref. [14] it was shown that this backscattered field can be written as:

$$\vec{E}_S = \frac{e^{-ik|\vec{r}|}}{4\pi|\vec{r}|} \vec{F}_S, \tag{14}$$

where complex vector $\vec{F}_S$ can be computed from edge element coefficients as the sum:

$$\vec{F}_S(\vec{e}_\rho) = -i\omega\mu \sum_{i=1}^{N}\sum_{j=1}^{3} \int_{\Delta_i} h_{bj} e^{ik\vec{r}\prime\cdot\vec{e}_\rho} d\vec{S}\prime \times \delta_j\vec{w}_j$$

$$+ ik \sum_{i=1}^{N}\sum_{j=1}^{3} \int_{\Delta_i} e_{bj} e^{ik\vec{r}\prime\cdot\vec{e}_\rho} \left(d\vec{S}\prime \times \delta_j\vec{w}_j\right) \times \vec{e}_\rho \qquad (15)$$

$$- ik \sum_{i=1}^{N}\sum_{j=1}^{3} \int_{\Delta_i} h_{bj} e^{ik\vec{r}\prime\cdot\vec{e}_\rho} dS\prime \frac{\nabla\prime_S \cdot (\vec{n}\prime \times \vec{w}_j)}{\sigma + i\omega\epsilon} \vec{e}_\rho.$$

Computation of vector function $\vec{F}_S(\vec{e}_\rho)$ from known boundary edge coefficients h_b and e_b is fast because there are no numerical computations of double surface integrals as in eqn (11) and because there are no hyper-singular integrals involved in the computation.

The method is tested for a number of different configurations of inhomogeneous scattering problems and some previously published results [12] of the comparison of the method with results obtained by commercial software FEKO are shown in Fig. 2. FEKO software uses hybrid MoM/FEM to calculate near field and from there it computes RCS using near-to-far field transformation (NTFFT). As shown at the bottom of Fig. 2, the plane wave was directed along negative z axis at composite cone-sphere object. The first top layer of the cone is dielectric with relative permittivity $\epsilon_r = 20$ and of height 0.2λ (where λ is the wavelength). Second layer of the cone is dielectric with relative permittivity $\epsilon_r = 2$ and of height 0.3λ. The cone is placed on perfect electric conductor (PEC) hemisphere of radius 0.5λ. The arrangement is difficult for solvers because there is sharp metallic tip where some problems at corners are to be expected. However, as shown in Fig. 2, FEKO and hybrid BEM/FEM described in this paper are in excellent agreement.

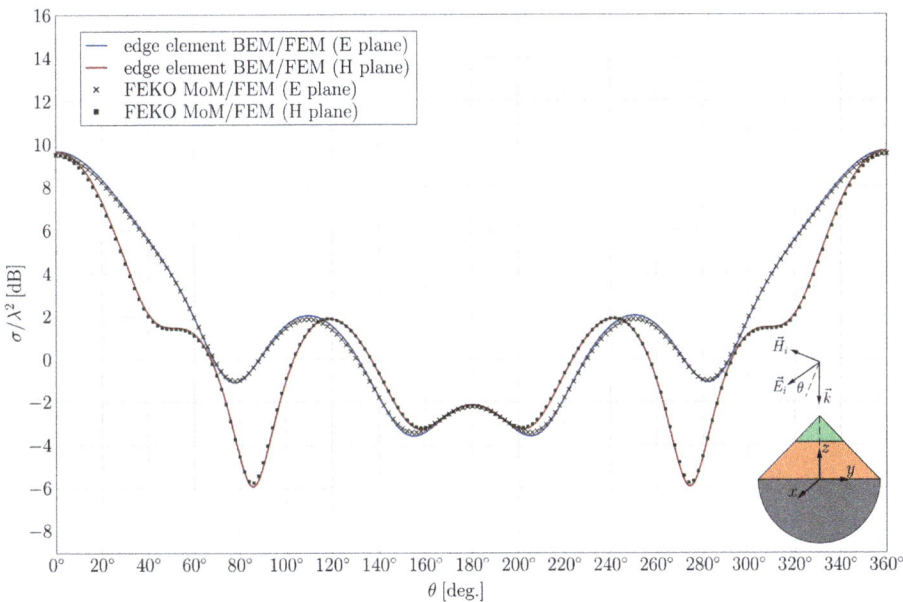

Figure 2:   Bistatic RCS of composite three layered sphere-cone object computed by hybrid BEM/FEM compared to results obtained by commercial software FEKO [11].

4 MODELS OF SMALL MARITIME OBJECTS

The rubber boat chosen for RCS computation is small service rubber boat used as a service boat in Croatian Navy. To correctly capture the geometry of the rubber boat it was subjected to series of 3D laser measurements as shown in Fig. 3. Laser measurement equipment was made available to authors by courtesy of Croatian Navy laser measurement facility site. Results of laser measurements were exported in the standard STL file format in order to produce nonconformal tetrahedral mesh of good quality the STL file was converted to Parasolid x_t CAD format using Geomagic software. The seawater shown in Fig. 4(a) into which the boat is submerged was added with Siemens SolidEdge CAD software. Nonconformal tetrahedral mesh shown in Fig. 4(b) was produced by Ansys ICEM software and it consists of 285,064 tetrahedral elements for FEM part of computation and of 3,632 triangular elements for BEM part of computation. Electrical parameters $\sigma, \epsilon_r, \mu_r$ of the seawater, air, rubber and plastic were compiled from various sources from literature [21] and these are given in Table 1.

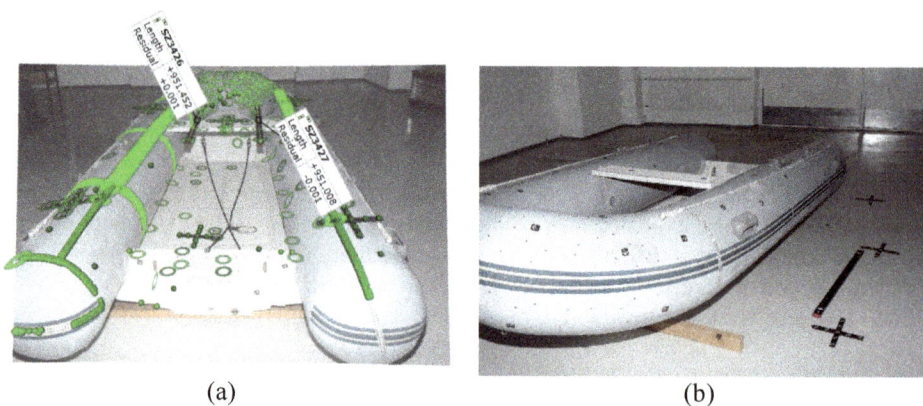

(a) (b)

Figure 3: Photo of the rubber boat at laser measurement site is displayed in (a). The geometric features of the boat are captured at spatial points marked with green dots. In (b) the preparation of laser measurement site is shown.

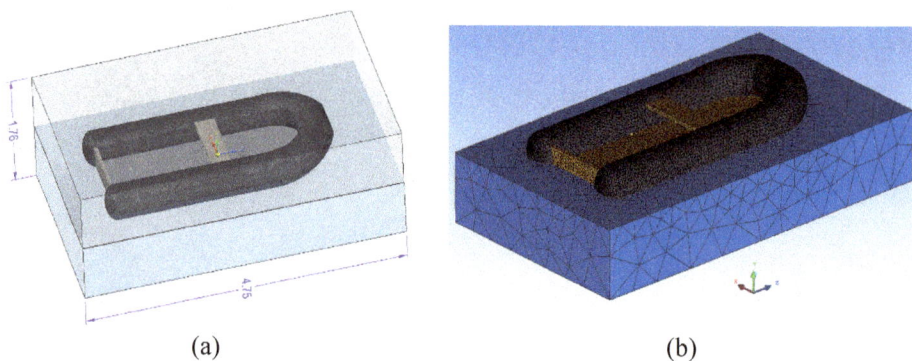

(a) (b)

Figure 4: CAD model of the geometry for RCS computation is shown in (a). The rubber boat is partially immersed in seawater. In (b) nonconformal tetrahedral mesh of the rubber boat is displayed.

Table 1: Electrical properties of materials for EM scattering calculation.

#	Electrical properties of materials		
	Material	σ [S/m]	ϵ_r
1.	Seawater	3.00	88.00
2.	Air	0.00	1.00
3.	Rubber	0.00	2.50
4.	Plastic	3.00	3.00
5.	Industrial steel	6.21×10^6	1.00

The CAD model of standard navigational buoy used along Croatian coastal area is depicted in Fig. 5(a). The model was built from the data made available by Croatian Hydrographic Institute (HHI). Nonconformal tetrahedral mesh of the buoy model and of surrounding seawater consisted of 986,260 tetrahedral elements and 1,870 elements for BEM surface.

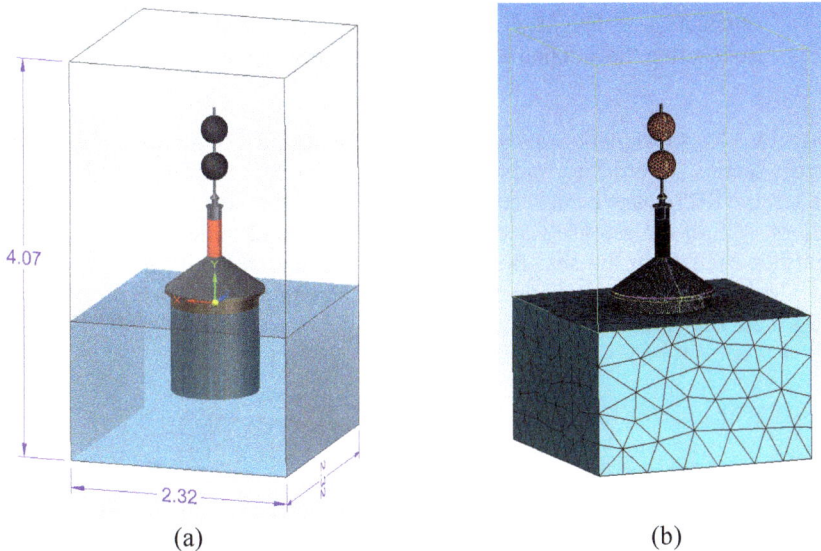

(a)   (b)

Figure 5:   CAD model of the navigational buoy partially submerged into seawater is displayed in (a). Nonconformal tetrahedral mesh of the buoy and seawater is shown in (b). Model consists of 986,260 tetrahedral elements and 1,870 triangular elements to model the computational boundary ∂V.

5 NUMERICAL RCS RESULTS

Physical setting for RCS computation of rubber boat and the buoy is displayed in Fig. 6. The rubber boat is at distance l from the ship and the radar antenna is located at height h above the sea level. In Fig. 6 angle θ represents azimuth, that is, the angle θ is the angle between the imaginary line connecting radar antenna and the rubber boat (buoy). Furthermore, RCS computations were performed for both horizontal polarization (HP) and vertical polarization

(VP). The direction of EM wave propagation is always along imaginary line connecting radar antenna and the rubber boat. For HP electric field is in the sea plane and for VP the magnetic field is in the sea plane.

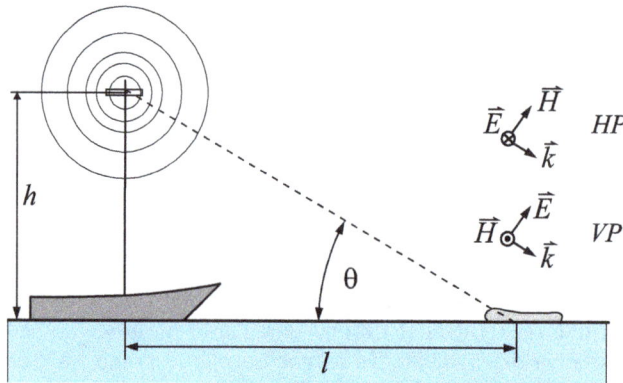

Figure 6: The rubber boat is at horizontal distance l from the radar antenna and the radar antenna is at height h above sea level. Angle subtended between the line connecting the antenna and boat and between sea level is θ.

Normally, electromagnetic waves reflect from the sea surface and even the sea itself has radar cross section. In order for radar to distinguish between the target (rubber boat, buoy) and the sea, the RCS has to be different. In fact, we could devise a measure corresponding to the ratio of RCS of the sea itself and the target which could indicate how much the target distinguishes itself from the sea. Because of this we have first calculated the radar cross section σ_S of the sea itself as a function of various azimuth angles θ, and the results for both HP and VP are shown in Fig. 7(a).

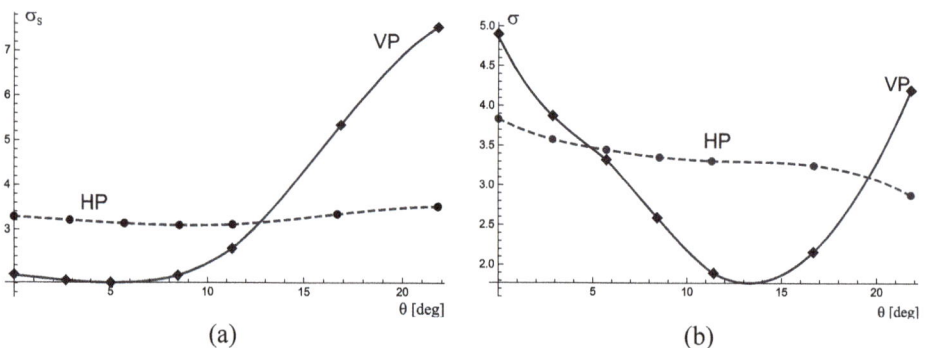

(a)   (b)

Figure 7: Radar cross section of the seawater (without rubber boat) for various incidence angles θ is shown in (a) for horizontal polarization (HP) and vertical polarization (VP). Radar cross section of both rubber boat and seawater is shown in (b) for HP and VP.

The calculations of radar cross section σ of rubber boat immersed in the sea, depicted in Fig. 4(b), is carried out next. Various azimuth angles are considered and the results for both HP and VP are displayed in Fig. 7(b). In order to find out how much radar cross section of the rubber boat distinguishes itself from the radar cross section of the sea the ratio σ/σ_S was calculated as shown in Fig. 8. When ratio σ/σ_S is close to 1 it is impossible for radar to distinguish the rubber boat from sea surface.

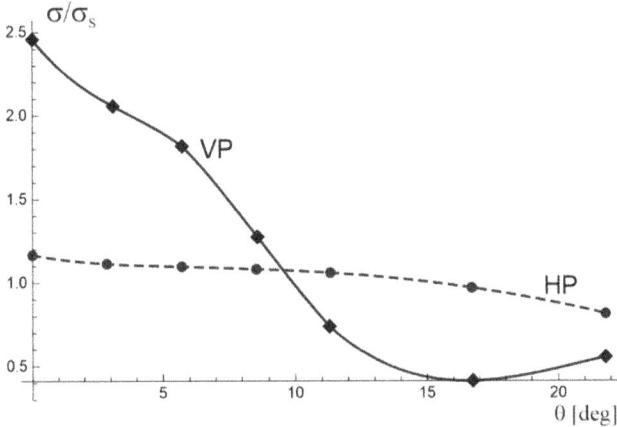

Figure 8:   The ratio σ/σ_S for rubber boat for both HP and VP polarization.

Similar calculations of the σ/σ_S ratio are performed for buoy for both HP and VP polarization and the results of this calculations are shown in Fig. 9.

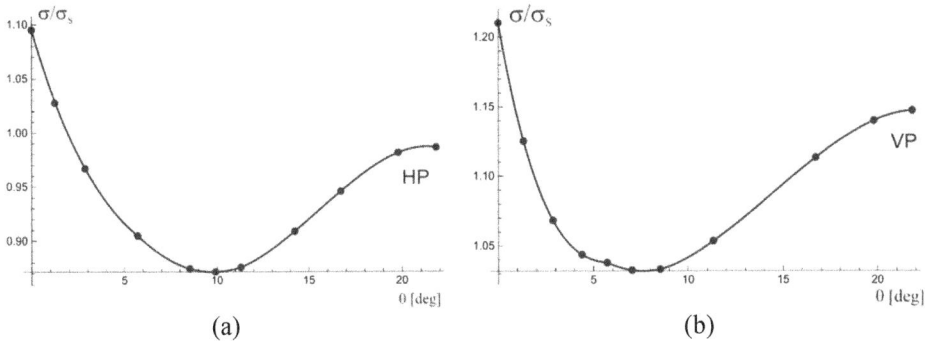

(a)   (b)

Figure 9:   The ratio of cross section σ/σ_S for buoy is shown for horizontal polarization in (a). In (b) the ratio σ/σ_S for buoy is shown for vertical polarization (VP).

The calculations of near electromagnetic fields for rubber boat and the buoy are shown in Fig. 10. From these field values RCS was calculated using eqn (15). The results shown in Figs 8–10 indicate that there are some azimuth angles θ where it is very difficult for radar to distinguish small target from the surface. Hence, further research will be carried out to investigate this possible safety issue.

WIT Transactions on Engineering Sciences, Vol 131, © 2021 WIT Press
www.witpress.com, ISSN 1743-3533 (on-line)

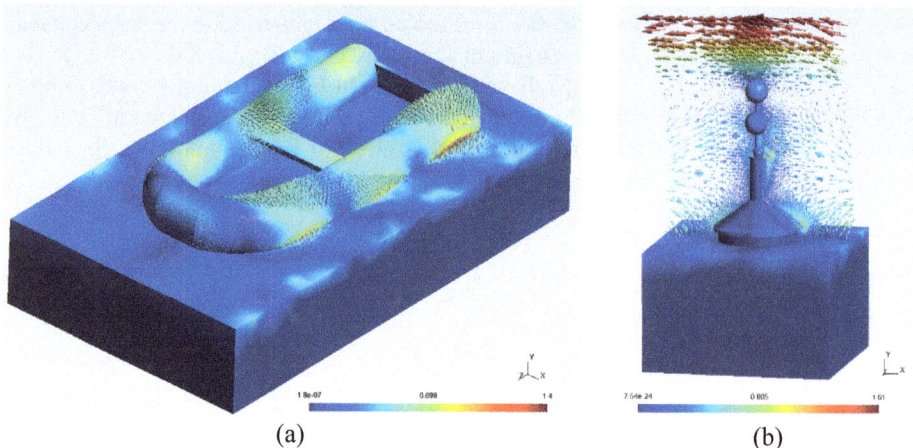

(a)   (b)

Figure 10: The vectors of near electric field are shown on the surface of rubber boat in (a), and the vectors of near electric field are shown in the air surrounding the navigational buoy in (b).

6 CONCLUSION

The present paper outlines an efficient method for RCS calculation of small maritime radar targets. The method presented is full-wave method that can take into account the material inhomogeneities of radar targets. The near field computation was performed using hybrid BEM/FEM with edge elements which bear certain advantages over RWG elements that are standardly used with competing MoM/FEM formulation: these elements are divergence-less (divergence free), hence, they naturally satisfy zero divergence condition and they preserve tangential continuity of electromagnetic fields across inter-element boundaries which is physically required condition. From near field values the RCS was computed using formulation based on edge elements. Namely, the results for RCS as a function of azimuth, are given for small maritime objects, i.e. small rubber boat and buoy. The RCS values for these small objects are compared to RCS of the seawater and some conclusions regarding detectability of small maritime objects by radar were drawn.

ACKNOWLEDGEMENT

We thank Mr. Ljubomir Ostavić from Croatian Ministry of Defense for performing the laser measurements of the service rubber boat.

REFERENCES

[1] Maini, A.K., *Handbook of Defence Electronics and Optronics: Fundamentals, Technologies and Systems*, John Wiley & Sons Ltd.: New York, pp. 113–202, 2018.
[2] Bole, A. et al., *Radar and ARPA Manual*, Elsevier: London, pp. 139–213.
[3] Knott, E.F., A progression of high-frequency RCS prediction techniques. *Proceedings of the IEEE*, **73**(2), pp. 252–264, 1985.
[4] Youssef, N.N., Radar cross section of complex targets. *Proceedings of the IEEE*, **77**(5), pp. 722–734, 1989.
[5] Silverstein, J.D. & Bender, R., Measurements and predictions of the RCS of Bruderhedrals at millimeter wavelengths. *IEEE Transactions on Antennas and Propagation*, **45**(7), pp. 1071–1079, 1997.

[6]   Weinmann, F., Ray tracing with PO/PTD for RCS modeling of large complex objects. *IEEE Transactions on Antennas and Propagation*, **54**(6), p. 1806, 2006.

[7]   Dodig, H. et al., Edge element calculation of radar cross section of small maritime targets with respect to height of radar antenna. *TransNav: International Journal on Marine Navigation and Safety of Sea Transportation*, **14**(2), pp. 331–336, 2020.

[8]   Liu, J. et al., An efficient ray-tracing method for RCS prediction in GRECO. *Microwave and Optical Technology Letters*, **55**, pp. 586–589, 2012.

[9]   Yee, K., Numerical solution of initial boundary value problems involving Maxwell's equations in isotropic media. *IEEE Transactions on Antennas and Propagation*, **14**(3), pp. 302–307, 1966.

[10] Xu, R.W. et al., A hybrid FEM/MoM technique for 3-D electromagnetic scattering from a dielectric object above a conductive rough surface. *IEEE Geoscience and Remote Sensing Letters*, **13**(3), pp. 314–318, 2016.

[11] Dodig, H. et al., On the computation of singular integrals over triangular surfaces in R3. *Boundary Elements and Other Mesh Reduction Methods XLI*, eds A. Cheng & S. Syngellakis, WIT Press: Southampton, UK, pp. 95–105, 2018.

[12] Dodig, H. et al., On the edge element boundary element method/finite element method coupling for time harmonic electromagnetic scattering problems. *International Journal for Numerical Methods in Engineering*, early view (online), pp. 1–40, 2021.

[13] Taflove, A. & Hagness, S.C., *Computational Electrodynamics: The Finite-Difference Time-Domain Method*, Artech House Inc.: Boston, pp. 329–352, 2005.

[14] Dodig, H., A boundary integral method for numerical computation of radar cross section of 3D targets using hybrid BEM/FEM with edge elements. *Journal of Computational Physics*, **348**, pp. 790–802, 2017.

[15] Dodig, H. et al., Hybrid BEM/FEM edge element computation of the thermal rise in the 3D model of the human eye induced by high frequency EM waves. *Proceedings of SOFTCOM*, 2012.

[16] Dodig, H. et al., Stochastic sensitivity of the electromagnetic distributions inside a human eye modeled with a 3D hybrid BEM/FEM edge element method. *Engineering Analysis with Boundary Elements*, **49**, pp. 48–62, 2014.

[17] Stratton, J. & Chu, L.J., Diffraction theory of electromagnetic waves. *Physical Review*, **56**, pp. 99–107, 1939.

[18] Harrington, R.F., *Time-Harmonic Electromagnetic Fields*, John Wiley & Sons: New York, p. 77, 2001.

[19] Nedelec, J.C., Mixed finite elements in R3. *Numerische Mathematik*, **35**, pp. 315–341, 1980.

[20] Cvetković, M. et al., A study on the use of compound and extracted models in the high frequency electromagnetic exposure assessment. *Mathematical Problems in Engineering*, web only/open access, **2017**, 7932604, 2017.

[21] Garazza, A.R.L. et al., Influence of the microstructure of vulcanized polybutadiene rubber on the dielectric properties. *Polymer Testing*, **30**, pp. 657–662, 2011.

COUPLED BOUNDARY ELEMENT: STOCHASTIC COLLOCATION APPROACH FOR THE UNCERTAINTY ESTIMATION OF SIMULATIONS OF TRANSCRANIAL ELECTRIC STIMULATION

JURE RAVNIK[1], ANNA ŠUŠNJARA[2], OŽBEJ VERHNJAK[1], DRAGAN POLJAK[2] & MARIO CVETKOVIĆ[2]
[1]Faculty of Mechanical Engineering, University of Maribor, Slovenia
[2]Faculty of Electrical Engineering, Mechanical Engineering and Naval Architecture, University of Split, Croatia

ABSTRACT

In this paper, the authors present a deterministic model of transcranial electric stimulation and develop a Boundary Element Method based algorithm capable of calculating potential and current density in the investigated domain. Furthermore, the deterministic model is coupled with Stochastic Collocation Method to evaluate the propagation of the uncertainty of the input parameters to the results. The uncertainty of the results is analysed via statistical approaches. The governing partial differential equation of the deterministic model is the Laplace equation. We assume that the studied domain is composed of subdomains of different materials. The subdomains are homogeneous and isotropic and have constant electrical conductivity. We introduce the Boundary Element Method for such a setup, support Dirichlet and Neumann boundary conditions at the outer boundary of the domain, and assume continuity of the potential and conservation of current density at the boundaries between the subdomains. We assume that the electrical conductivity of each subdomain is subject to some uncertainty. By coupling the developed simulation tool with the Stochastic Collocation Method, we are able to analyse the uncertainty of the resulting electric potential and current density and identify the contribution to the uncertainty from each subdomain. We apply the developed algorithms to study the transcranial electric stimulation of a human head. A head model with 9 tissues (white and grey matter parts of cerebellum, ventricles, cerebellum, cerebrospinal fluid, head, tongue, cerebrum, and skull) is considered and voltage is applied across two electrodes. The results show that the uncertainty of electrical conductivity in the skull, cerebrospinal fluid and, grey matter have the largest influence on the results.
Keywords: boundary element method, stochastic collocation method, sensitivity analysis, transcranial electric stimulation.

1 INTRODUCTION

The Boundary Element Method (BEM) is a numerical technique aimed at solving partial differential equations (PDEs). In the case of a homogenous PDE, using its fundamental solution, we derive a boundary integral equation, which enables us to find the solution by discretizing only the boundary of the problem domain. The authors consider the modelling of Transcranial Electric Stimulation (TES) in this paper. Transcranial electrical stimulation (TES) represents a non-invasive brain stimulation technique used as a treatment in various brain related disorders, as well as an important tool in stroke recovery and chronic pain [1]. Crucial to the application of TES is the knowledge on the conductivity of biological tissues. However, mostly due to a different measurement methods and tissue preparation techniques, but also due to particularly challenging measurement in the low frequency range, there is a high level of discrepancy between the reported values of tissue properties. These tissue uncertainties are considered a main concern in computational models of TES since tissue conductivity values will have a significant impact on the distribution of the induced fields [1]. Therefore, we consider the coupling of deterministic TES model with statistical methods to take into account this implicit uncertainty. TES, for a homogenous domain, is governed by

WIT Transactions on Engineering Sciences, Vol 131, © 2021 WIT Press
www.witpress.com, ISSN 1743-3533 (on-line)
doi:10.2495/BE440121

the Laplace equation. When applied to the stimulation of the human head, inhomogeneous conditions as the electrical conductivity of different tissues varies are encountered. Domain decomposition (Bui et al. [2]) is considered to handle different conductivities of the different tissues.

The main goal of this work is to couple the deterministic BEM based simulation tool with the Stochastic Collocation Method (SCM, Babuska et al. [3]) to estimate the uncertainty of the numerical predictions. Numerical modelling and simulation are advantageous in many cases to performing experimental measurements and are thus widely used in almost all engineering disciplines for design and optimization purposes. The use of a purely deterministic model and a BEM solution leads to simulation results, but does not contain any information about the uncertainty of the results. Such an estimation can be obtained by comparing simulation results with experiments, but these are costly and time consuming. The SCM allows the propagation of the uncertainty of input parameters (such as tissue conductivity in the case of TES) through the numerical model to the results (such as electric potential or current in the case of TES).

2 GOVERNING EQUATIONS

To study bioelectromagnetism in living tissues, the authors consider them as volume conductors [4]. Their resistances, capacitances, and voltage sources are distributed within a volume, and the inductive component of the impedance is neglected. For low frequencies, we can neglect capacitive and electromagnetic effects. Thus, currents and voltages can be considered stationary within this quasi-static approximation.

Let us consider a tissue as a conductor with electrical conductivity σ in which there are no volumetric current sources. If an electric field $\vec{E}$ is assumed, the net current flow into/out of the volume must be solenoidal, i.e.

$$\vec{\nabla} \cdot \vec{J} = 0, \tag{1}$$

where $\vec{J}$ is the current density. It is related to the electric field via $\vec{J} = \sigma \vec{E}$. In static conditions, the electric field may be expressed as a negative gradient of the electric potential φ, $\vec{E} = -\vec{\nabla}\varphi$. With this, we may rewrite eqn (1) as

$$\vec{\nabla} \cdot \left(-\sigma \vec{\nabla}\varphi\right) = 0. \tag{2}$$

In general, the electrical conductivity for an anisotropic conductor is a tensor. For example, in the white matter of the brain, the electrical conductivity is higher in the direction of the nerve fiber tracts. In this work, we model the tissues in the human head as a group of subdomains, each representing individual tissues with different but homogeneous and isotropic electrical conductivity. In this case, the eqn (2) for each subdomain simplifies to a Laplace equation

$$\nabla^2 \varphi = 0, \tag{3}$$

where changes of electric conductivity between tissues are taken into account through boundary conditions. At the outer boundary of the computational domain we prescribe either Dirichlet boundary condition (known electric potential φ) or Neumann boundary conditions (known projection of electric current density to normal direction $J = -\sigma \vec{n} \cdot \vec{\nabla}\varphi$). Here, $\vec{n}$ is the outward pointing unit normal at the boundary. At the boundaries between subdomains continuity of potential ($\varphi_I = \varphi_{II}$) and conservation of current density ($J_I = -J_{II}$) is prescribed.

WIT Transactions on Engineering Sciences, Vol 131, © 2021 WIT Press
www.witpress.com, ISSN 1743-3533 (on-line)

3 NUMERICAL ALGORITHM

3.1 The boundary element method for TES

A three-dimensional domain $\Omega \in \mathbb{R}^3$ with a boundary $\Gamma = \partial\Omega$ and a location vector $\vec{r}$ is considered. The domain is locally homogeneous, so different sub-regions have different, but constant material properties. The authors employ the domain-decomposition approach and divide the domain into subdomains $\Omega = \cup\Omega_i$, where each subdomain has its own boundary $\Gamma_i = \partial\Omega_i$. Inside each subdomain we can write a boundary integral representation of the Laplace eqn (3) as [5]

$$c(\vec{\xi})\varphi(\vec{\xi}) + \int_{\Gamma_i} \varphi(\vec{r})\vec{\nabla}\varphi^\star \cdot \vec{n}d\Gamma_i = \int_{\Gamma_i} \varphi^\star(\vec{n} \cdot \vec{\nabla}\varphi(\vec{r}))d\Gamma_i \qquad \vec{\xi} \in \Gamma_i, \qquad (4)$$

where Γ_i is the boundary of ith subdomain, $\vec{\xi}$ is the source point, c is the free coefficient, and $\varphi^\star = 1/4\pi|\vec{r} - \vec{\xi}|$ is the fundamental solution of the Laplace operator. Such a representation allows us to solve only for the unknowns at the boundary of the subdomain, since the solution in the interior depends only on the knowledge of boundary variables (potential $\varphi(\vec{r})$ and flux $q = \vec{n} \cdot \vec{\nabla}\varphi(\vec{r})$).

To obtain a system of linear equations for the unknowns at the boundary, we discretize subdomain boundaries with triangular elements. Within the triangles we use a linear interpolation of the potential $\varphi(\vec{r}) = \sum \Phi_j\varphi(\vec{r})_j$ and a constant interpolation of the flux. The collocation point is placed into each interpolation vertex (the three corners and the barycentre of each boundary element). The two integrals in eqn (4) are evaluated and stored into matrices $[H_i]$ and $[G_i]$. Each row of the matrix corresponds to a collocation point and each column to a node in the computational mesh. The matrix entries are

$$[H_i^{(jk)}] = \int_{\Gamma_i^{(k)}} \Phi_k\vec{\nabla}\varphi_j^\star \cdot \vec{n} \, d\Gamma, \qquad [G_i^{(jk)}] = \int_{\Gamma_i^{(k)}} \varphi_j^\star d\Gamma. \qquad (5)$$

where i denotes the subdomain, $\Gamma_i^{(k)}$ is the boundary element, j the collocation point and k the column index corresponding to the node index of the chosen boundary element and interpolation shape function. Contribution to a node k from different adjacent boundary elements, which share the node k are summed up and stored in a single matrix entry. Using the matrices of integrals we obtain the following system of linear equations for each subdomain

$$c_i\varphi_i + [H_i]\{\varphi\} = [G_i]\{q\}, \qquad (6)$$

where $\{\varphi\}$ and $\{q\}$ are vectors of nodal values of potential and flux. The computation of the free coefficient $c(\vec{\xi})$ and the strongly singular diagonal element of $[H_i]$ is done indirectly. Setting $\varphi = 1$, $q = 0$ as one of the valid solutions of the original problem, we can use eqn (6) to evaluate the sum of c and the diagonal element in the $[H_i]$ matrix if all other elements were previously evaluated by numerical integration.

Integrals in eqn (5) are performed numerically over triangles in 3D space. For the purpose of linear interpolation within each triangle it is convenient to use barycentric coordinate system $(\lambda_1, \lambda_2, \lambda_3)$. A position within a triangle is then calculated by $\vec{r} = \lambda_1\vec{r}_1 + \lambda_2\vec{r}_2 + \lambda_3\vec{r}_3$. Weights and integration point locations were taken out of [6] and converted to the barycentric coordinate system. Lists of $(w_i, \lambda_{1,i}, \lambda_{2,i}, \lambda_{3,i})$ with 7, 25, 54, 85 and 126 entries were obtained, which provide machine precision accurate integration of polynomials of degree 5, 10, 15, 20 and 25 respectively. With this, integral over a triangular boundary element

τ can be approximated by

$$\int_\tau f(\vec{r})d\Gamma \approx \sum_i w_i f(\lambda_{1,i}\vec{r}_1 + \lambda_{2,i}\vec{r}_2 + \lambda_{3,i}\vec{r}_3). \tag{7}$$

When applying the boundary element method singular integrals must be computed. The singularity is located at one of the vertices of the triangle (in linear interpolation methods) or at the centre of the triangle (in constant interpolation methods). To accurately compute such integrals, we recursively subdivide the triangle in the direction of the singular point. Fig. 1 demonstrates the recursive algorithm. At each step, the triangle is subdivided into four smaller triangles obtained by bisecting the sides of the triangle. The one small triangle that still contains the singularity is subdivided again in the next step. The number of parts into which the triangle is divided is $1 + 3n$, where n is the number of recursive steps. The final result is a sum of integrals calculated over all the parts. The number of recursive steps controls the accuracy and can be adjusted to match the accuracy of computing non-singular integrals.

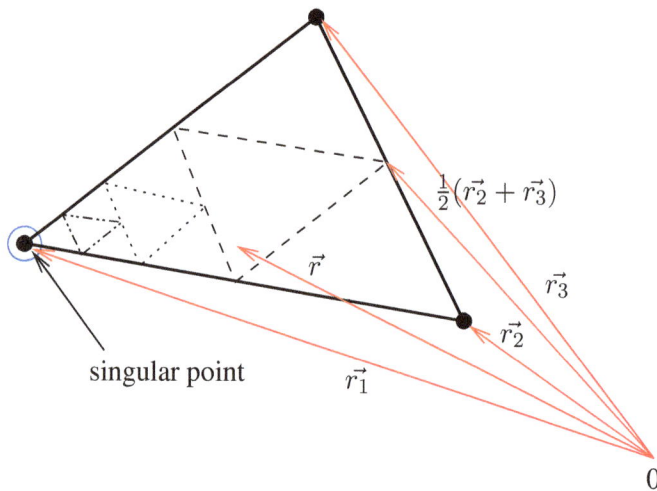

Figure 1: Recursive subdivision of a triangle element towards the singular point.

Having set up the systems of linear equations (eqn (6)) for each subdomain, let us examine them and discuss the boundary conditions. The number of equations (rows in matrices) in these systems is equal to the number of all nodes in the mesh. This includes all vertices of all triangles and all barycenters. For each boundary element located at the outer boundary, either the potential (at the triangle vertices) or the flux (at the barycenter) is known. At parts of the boundary, which are adjacent to neighboring subdomains, the potential and flux are unknown. Such a system is underdetermined and cannot be solved alone. To mitigate this, continuity of potential and conservation of current density at the boundaries between subdomains is assumed. This gives us an overdetermined system of linear equations (since the nodes are shared between subdomains). The resulting overdetermined linear system of equations is solved for the unknown potential and flux at outer and inner boundary using a least squares based solver (Paige and Saunders [7]). An alternative approach was proposed by Loeffler and Mansur [8], who use a domain superposition instead of a domain decomposition. After the solution at the boundaries has been found, it is possible to acquire the value of potential at any point in the domain explicitly by using eqn (4) and setting the source point $\vec{\xi}$ at the desired location.

3.2 Numerical code verification

To verify the numerical algorithm consider a 1D potential distribution problem within three joined subdomains. The domain is a long cylinder of diameter 1 m and length 3 m, as shown in Fig. 2. The prescribed potential is at the top and bottom of the cylinder is $\varphi_0 = 1$ V, $\varphi_3 = 4$ V and the conductivities are $\sigma_1 = 2$ S/m, $\sigma_2 = 3$ S/m, $\sigma_3 = 4$ S/m. There is no current at the cylinder wall, so the problem is one-dimensional. The analytical solution for the potential between the subdomains is $\varphi_1 = 34/13$ V, $\varphi_2 = 66/39$ V and for the current density is $J = 36/13$ A/m^2. Several discretizations of the domain were prepared and the results were compared with analytical values. Relative error is used to present the results in Fig. 3. We observe the developed algorithm is second order accurate for current density and even better for potential. At accuracy below 10^{-6} convergence stops due to the finite accuracy of numerical integration. In absolute sense the accuracy of potential is better than the accuracy of current density due to the fact we use linear interpolation of potential over boundary elements and only constant interpolation of current density within the boundary elements.

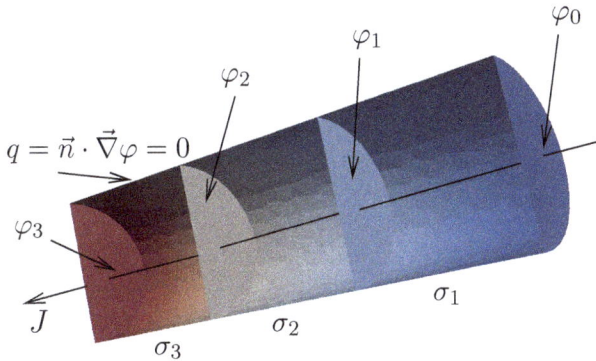

Figure 2: Potential distribution in the case of one-dimensional electrostatics problem with three subdomains. Constant potential φ_0 and φ_3 are prescribed at the top and bottom ends of the cylinder, whereas zero flux is assumed at the cylinder surface.

4 STOCHASTIC COLLOCATION METHOD

We consider the conductivity of the tissues in head model to be random variables uniformly distributed in a range $\sigma \in (\sigma_{min}, \sigma_{max})$. The size of this interval is due to changes between individuals, which was encountered when measuring samples of brain tissue from different donors. We assume uniform distribution, so probability distribution function (PDF) of all random variables is non-zero only in this range: $(\sigma_{min}, \sigma_{max})$:

$$p(\sigma) = \begin{cases} \frac{1}{\sigma_{max} - \sigma_{min}} & \sigma \in (\sigma_{min}, \sigma_{max}). \\ 0 & elsewhere \end{cases} \tag{8}$$

Let the number of random variables (different tissue conductivities in our model) be n. Additionally let our deterministic model, which solves for the potential in the head and

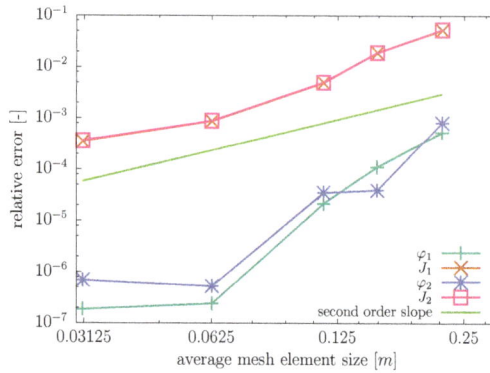

Figure 3: Relative error in potential and current density calculated as the difference between analytical and numerical solution at interfaces between subdomain. The green line shows the second order convergence slope. Average mesh element size is calculated as a square root of average mesh element area.

depends on the conductivities, be denoted by $y(\sigma_1, \ldots, \sigma_n)$. In this case, statistics for our deterministic model such as expected value μ_y and variance var_y may be calculated using [9]:

$$Y_i = \int_{-\infty}^{\infty} \cdots \int_{-\infty}^{\infty} [y(\sigma_1, \ldots, \sigma_n)]^i \, p(\sigma_1) \ldots p(\sigma_n) d\sigma_1 \ldots d\sigma_n, \tag{9}$$

$$\mu_y = Y_1, \qquad var_y = Y_2 - \mu_y^2. \tag{10}$$

Evaluating the integral (9) using a standard approach such as Gauss–Lengendre quadrature is possible, however it is very CPU intensive, as the number of evaluations of the model y scales as N^n, where N is the number of quadrature sample points. To avoid this, we use the Smolyak [10]–[12] sparse grid approach to numerically evaluate the integral (9). The integral is approximated by

$$Y_i \approx \frac{1}{2^n} \sum_{i=1}^{N_s} \left[y\left(\xi_i^{(1)}, \xi_i^{(2)}, \ldots, \xi_i^{(n)}\right) \right]^i w_i, \tag{11}$$

where $\xi_i^{(t)} = \sigma_{t,min} + (\sigma_{t,max} - \sigma_{t,min})\frac{\eta_i+1}{2}$ and η_i and w_i are sparse grid points and weights and the $\frac{1}{2^n}$ factor comes as a results of change of variables. For large n the number of sparse grid points in eqn (11) and with this the number of evaluations of the deterministic model is much smaller than the number of points needed by standard approaches, such as the Gauss–Lengendre quadrature, i.e. $N_s \ll N^n$. To calculate the points and weights for the sparse grid we employed the Tasmanian library [13], [14] using Clenshaw Curtis fully nested points.

5 RESULTS

We apply the developed method to a human head model with nine tissues. The model and the considered tissues are shown in Fig. 4. Since the ventricles are filled with cerebrospinal fluid, we consider the conductivity of these two subregions as one parameter. The indeterminacy of the conductivity in the jaw was neglected because it has several orders of magnitude less influence on the results compared to the other conductivities. Thus, a total of 7 conductivities are considered as random variables: Cerebellum, CSF, grey matter, head, tongue, cerebrum and skull. The potential at the electrodes was set to ± 1 V. Zero current boundary condition was used at the outer surface of the head.

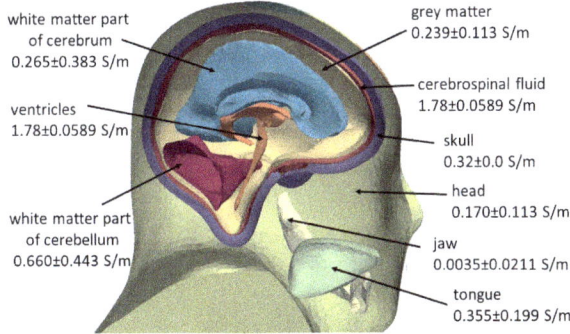

Figure 4:   The nine tissue head model, with subdomains representing ventricles, white matter parts of cerebrum and cerebellum, grey matter, cerebrospinal fluid, skull, jaw and tongue. All tissues are enclosed into a domain named *head*, which represents all other tissues in the human head. Conductivity values are obtained from the tissue properties database [15]. Ventricles are filled with cerebrospinal fluid, thus the conductivity of these two subdomains is treated as one parameter.

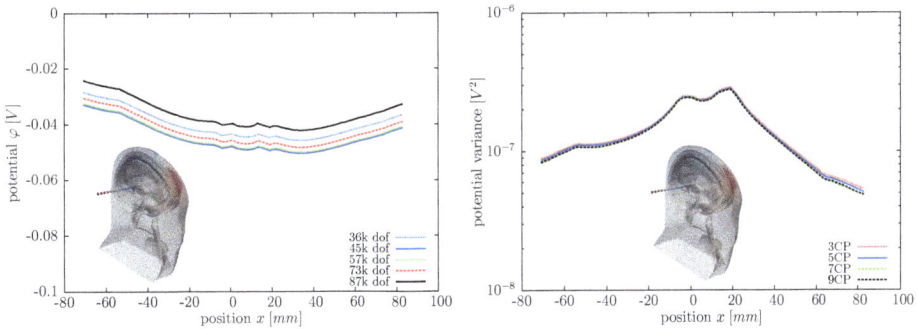

Figure 5:   Comparison of potential at left-to-right profile through the head model. Left: Difference in potential due to computational grid density in BEM. Right: Difference in potential variance due to difference in the number of collocation points in SCM. Only small differences between meshes and the number of collocation points are observed.

Different computational grids were constructed with 38,000 to 87,000 degrees of freedom. Only small differences in the results were observed (Fig. 5) so in the interest of saving computational time, we chose the grid with 45,000 degrees of freedom for the analysis. Furthermore, we used SCM with 3, 5, 7 and 9 collocation points to make sure, that the numerical calculation of the integral (9) converges. Comparison in potential variance is shown in Fig. 5 and reveals good agreement between results and proves that the numerical calculation of the integral converged. Finally, we chose Smolyak sparse grid approach with 589 simulations to be performed. The total CPU time on a single processor used was about 5 hours per simulation, in total 123 days.

In Fig. 6 we plot the variance of the scalar electric potential $[V^2]$ at the surfaces of the different tissues in the head. Overall, the largest variance $\approx 10^{-3}$ V^2 is observed in the white matter, grey matter, CFS, skull and head. It makes sense for the CSF to have the large variance due to high conductivity. It is considered a so called "super highway" for current flow, according to Bikson et al. [16]. The greatest variance is observed in the area under the electrodes. This analysis provides information about the uncertainty of the applied electric field to different tissues due to the unknown structure and properties of the tissues.

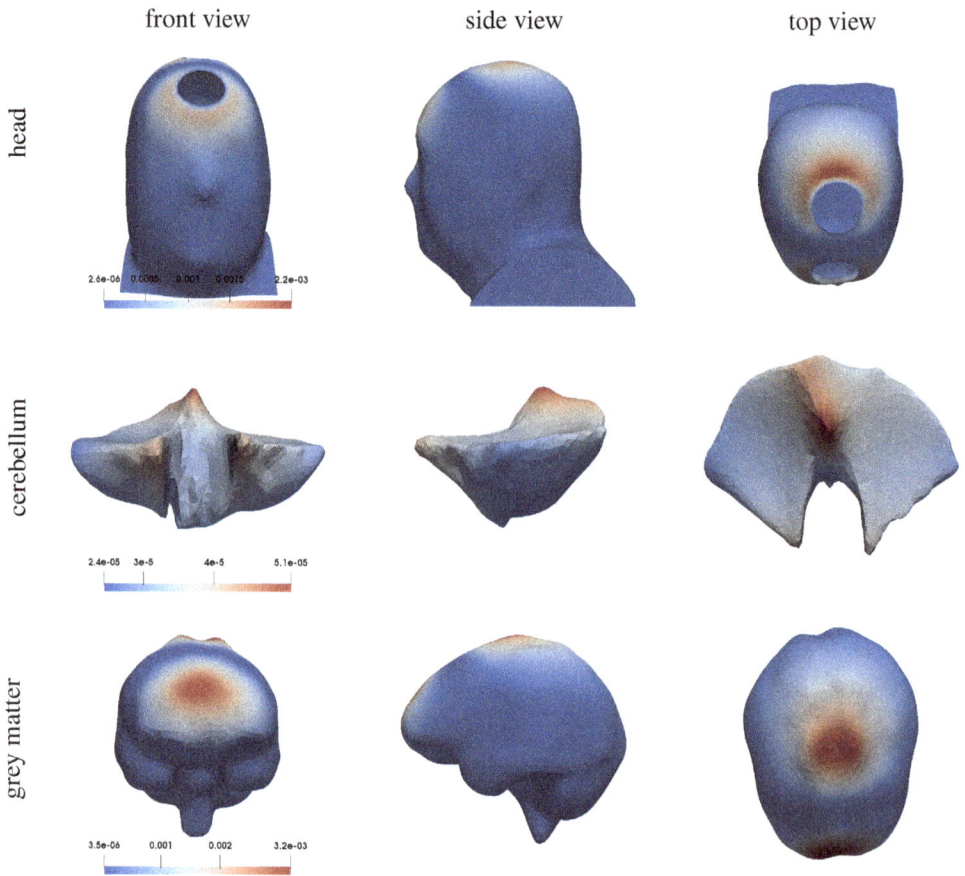

Figure 6: Variance of scalar electric potential $[V^2]$ shown at surfaces of different tissues in the head.

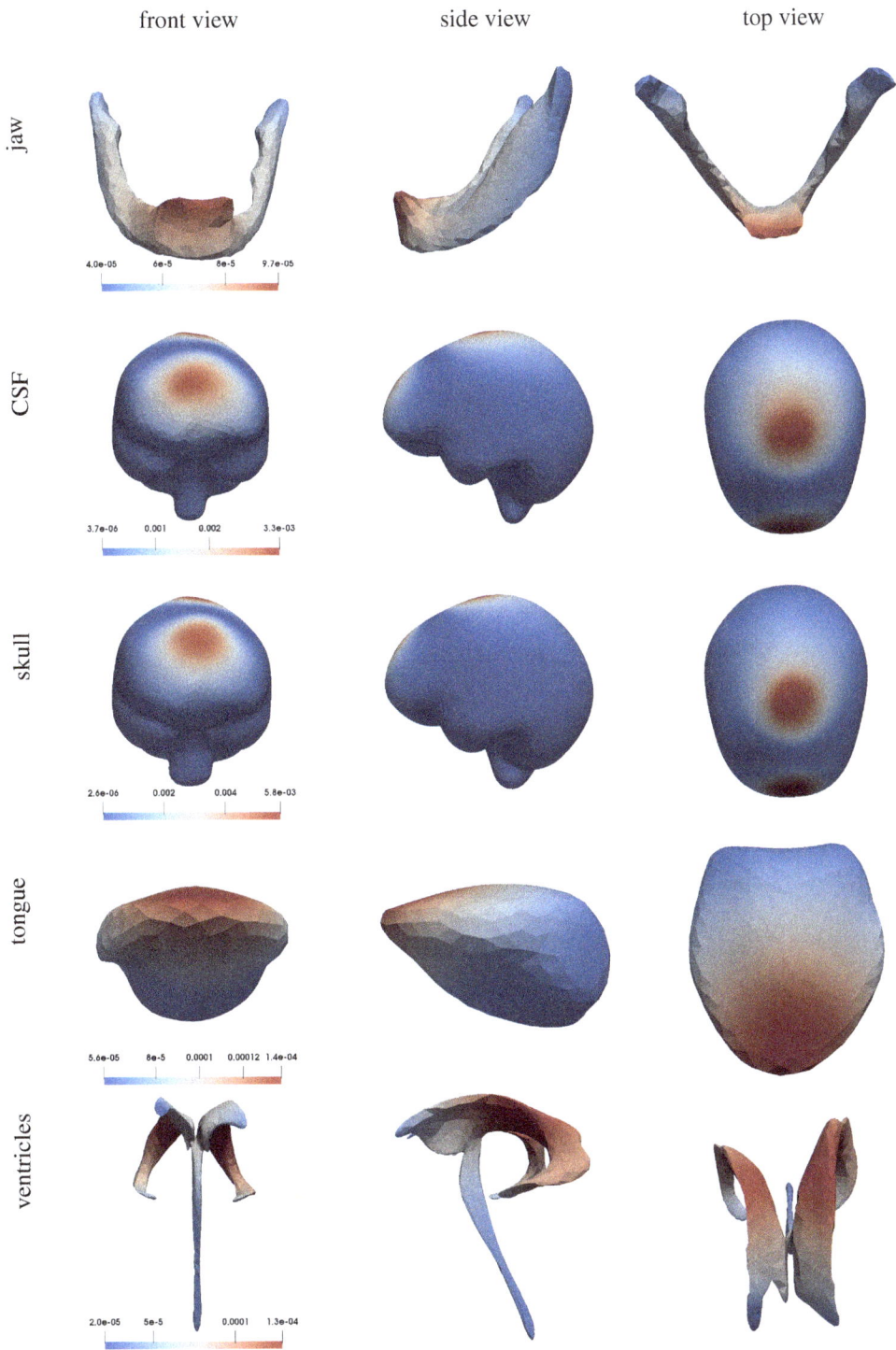

Figure 6: Continued.

front view   side view   top view

white matter

5.0e-06 0.0005 0.001 0.0015 2.1e-03

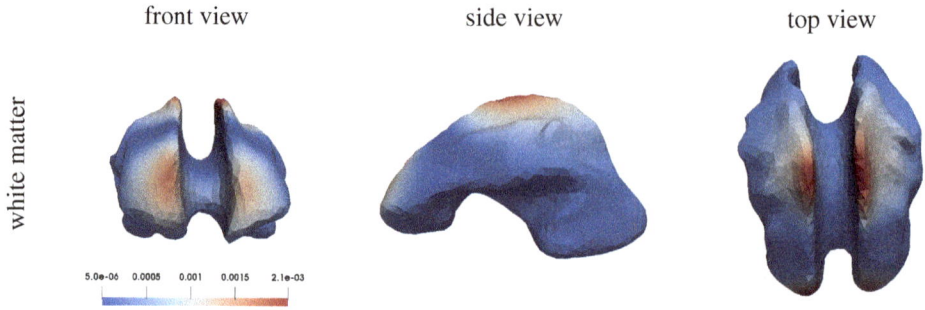

Figure 6: Continued.

6 CONCLUSIONS

In this paper, we introduce the Boundary Element Method to simulate the Transcranial Electric Stimulation and couple it with the Stochastic Collocation Method. BEM is a deterministic solver and the SCM wraps it to evaluate the uncertainty quantification of the output of interest. A domain decomposition technique is proposed to deal with changes in material properties in different regions of the model. The domain decomposition technique employed leads to a solution of an overdetermined system of linear equations. The SCM was implemented using a sparse numerical integration scheme, which greatly reduces the number of required deterministic simulations and makes such analyses possible for complex geometries.

The developed method is used to study the TES of a human head model with 9 tissues. The tissues are assumed to be homogeneous and isotropic with constant conductivities. The conductivities of the tissues are considered as random variables uniformly distributed in ranges obtained from the available literature. The expected value and variance values of the electrical scalar potential in the head tissues are calculated. This analysis reveals areas in the human head where the uncertainty of the applied electrical potential is greatest during the TES procedure. We show that the areas directly under the electrodes have the largest variance of the potential. Large variance is observed in white matter, grey matter, cerebrospinal fluid, skull and head.

REFERENCES

[1] Reato, D., Salvador, R., Bikson, M., Opitz, A., Dmochowski, J. & Miranda, P.C., *Principles of Transcranial Direct Current Stimulation (tDCS): Introduction to the Biophysics of tDCS*, Springer International Publishing: Cham, pp. 45–80, 2019.
[2] Bui, T.T. & Popov, V., Domain decomposition boundary element method with overlapping sub-domains. *Engineering Analysis with Boundary Elements*, **33**(4), pp. 456–466, 2009.
[3] Babuska, I., Nobile, F. & Tempone, R., A stochastic collocation method for elliptic partial differential equations with random input data. *SIAM Review*, **52**(2), pp. 317–355, 2010.
[4] Bai, S., Loo, C. & Dokos, S., A review of computational models of transcranial electrical stimulation. *Critical Reviews in Biomedical Engineering*, **41**(1), pp. 21–35, 2013.
[5] Wrobel, L.C., *The Boundary Element Method*, John Willey & Sons, Ltd, 2002.
[6] Wandzurat, S. & Xiao, H., Symmetric quadrature rules on a triangle. *Computers and Mathematics with Applications*, **45**(12), pp. 1829–1840, 2003.

[7] Paige, C.C. & Saunders, M.A., LSQR: An algorithm for sparse linear equations and sparse least squares. *ACM Transactions on Mathematical Software*, **8**, pp. 43–71, 1982.

[8] Loeffler, C.F. & Mansur, W.J., Sub-regions without subdomain partition with boundary elements. *Engineering Analysis with Boundary Elements*, **71**, pp. 169–173, 2016.

[9] Šušnjara, A., Verhnjak, O., Poljak, D., Cvetković, M. & Ravnik, J., Stochastic-deterministic boundary element modelling of transcranial electric stimulation using a three layer head model. *Engineering Analysis with Boundary Elements*, **123**, pp. 70–83, 2021.

[10] Smolyak, S., Quadrature and interpolation formulas for tensor products of certain classes of functions. *Soviet Mathematics Doklady*, **4**, pp. 240–243, 1963.

[11] Barthelmann, V., Novak, E. & Ritter, K., High dimensional polynomial interpolation on sparse grids. *Advances in Computational Mathematics*, **12**(4), pp. 273–288, 2000.

[12] Nobile, F., Tempone, R. & Webster, C.G., A sparse grid stochastic collocation method for partial differential equations with random input data. *SIAM Journal on Numerical Analysis*, **46**(5), pp. 2309–2345, 2008.

[13] Stoyanov, M., User manual: TASMANIAN sparse grids. Technical Report ORNL/TM-2015/596, Oak Ridge National Laboratory, Oak Ridge, TN, 2015.

[14] Stoyanov, M.K. & Webster, C.G., A dynamically adaptive sparse grids method for quasi-optimal interpolation of multidimensional functions. *Computers & Mathematics with Applications*, **71**(11), pp. 2449–2465, 2016.

[15] ITIS, Tissue properties database, 2019.

[16] Bikson, M., Asif, R. & Datta, A., Computational models of transcranial direct current stimulation. *Clinical EEG and Neuroscience*, **43**(3), pp. 176–183, 2012.

Author index

WIT*PRESS* ...for scientists by scientists

Excel in Complex Variables with the Complex Variable Boundary Element Method

*Edited by: **B.D. WILKINS**, Carnegie Mellon University, USA and **T. V. HROMADKA II**, United States Military Academy, USA*

Using the familiar software Microsoft ® Excel, this book examines applications of complex variables. Implementation of the included problems in Excel eliminates the "black box" nature of more advanced computer software and programming languages and, therefore, the reader has the chance to become more familiar with the underlying mathematics of the complex variable problems.

This book consists of two parts. In Part I, several topics are covered that one would expect to find in an introductory text on complex variables. These topics include an overview of complex numbers, functions of a complex variable, and the Cauchy integral formula. In particular, attention is given to the study of analytic complex variable functions. This attention is warranted because of the property that the real and imaginary parts of an analytic complex variable function can be used to solve the Laplace partial differential equation (PDE). Laplace's equation is ubiquitous throughout science and engineering as it can be used to model the steady-state conditions of several important transport processes including heat transfer, soil-water flow, electrostatics, and ideal fluid flow, among others.

In Part II, a specialty application of complex variables known as the Complex Variable Boundary Element Method (CVBEM) is examined. The CVBEM is a numerical method used for solving boundary value problems governed by Laplace's equation. This part contains a detailed description of the CVBEM and a guide through each step of constructing two CVBEM programs in Excel. The writing of these programs is the culminating event of the book.

Students of complex variables and anyone with interest in a novel method for approximating potential functions using the principles of complex variables are the intended audience for this book. The Microsoft Excel applications (including simple programs as well as the CVBEM program) covered will also be of interest in the industry, as these programs are accessible to anybody with Microsoft Office.

ISBN: 978-1-78466-451-0   **eISBN: 978-1-78466-452-7**
Published 2021 / 290pp

www.ingramcontent.com/pod-product-compliance
Lightning Source LLC
Chambersburg PA
CBHW062008190326
41458CB00009B/3005